Third Edition

Introductory Physical Science

Third Edition

Introductory

Physical Science

Haber-Schaim / Judson B. Cross / Gerald L. Abegg / John H. Dodge / James A. Walter

ntice-Hall, Inc., Englewood Cliffs, New Jersey

Introductory Physical Science

Third Edition

Uri Haber-Schaim / Judson B. Cross / Gerald L. Abegg / John H. Dodge / James A. Walter

Printed in the United States of America.

ISBN 0-13-502997-X cloth

ISBN 0-13-502989-9 paper

10 9 8 7

Prentice-Hall International, Inc., *London*
Prentice-Hall of Australia, Pty. Ltd., *Sydney*
Prentice-Hall of Canada, Ltd., *Toronto*
Prentice-Hall of India Private Ltd., *New Delhi*
Prentice-Hall of Japan, Inc., *Tokyo*

Preface to the Third Edition

This is a year-long course in introductory physical science. Its purpose is to give all students a beginning knowledge of physical science and to offer some insight into the means by which scientific knowledge is acquired. The course is designed to serve as a solid foundation for those students taking later courses in biology, chemistry, and physics and for those taking no further science.

The course not only helps students to understand some of the basic principles of physical science and to acquire useful laboratory skills, but also provides the opportunity for the development of communication skills within the context of science. This is done through the format of the course which encourages the individual student to take part in a collective learning process. The IPS students learn from nature, the text, their teacher, and each other.

The theme of the course is the development of evidence for an atomic model of matter. To this end we have taken a well-defined path toward this major objective rather than broadly surveying the entire field of physical science. The method employed to achieve the stated goals is one of student experimentation and guided reasoning on the results of such experimentation. Thus the laboratory experiments are contained in the body of the text and must be carried out by the students for the proper understanding of the course. Many of the conclusions and generalizations arrived at as a result of doing the experiments become essential parts of the complete text.

Although laboratory space is always an asset, the experiments in this course have been successfully performed in classrooms containing individual flat desk tops and only one sink.

IPS is now entering its second decade as a widely used program in introductory physical science. Indeed, it is most gratifying that thousands of science teachers across the country continue to find IPS an inspiration to their students. It will come as no surprise, therefore, that the basic content and philosophy of the program are carried over into the Third Edition. Nevertheless, important improvements were made in this edition in response to suggestions from experienced teachers.

We have shortened Chapter 3, Characteristic Properties, by concentrating only on freezing point, boiling point, and density (in this order). The compressibility of gases, and the thermal expansion of gases, liquids, and solids have been moved to Chapter 9, Molecular Motion. This change was made to enable slow classes to get further into the atomic model than they were able to do before.

The introductory sections of Chapter 4, Solubility, have been rewritten and a new experiment added. The experiment on fractional crystallization has been simplified.

Most of Chapter 7 has been eliminated, leaving just enough radioactivity to lead into the atomic nature of matter.

Since only the chapter on the sizes and masses of atoms and molecules uses exponential notation to any large degree, we have placed it at the end of the book.

While we hope to interest new teachers in IPS, we have not forgotten the many teachers who have been teaching it for years: about one fourth of the problems and HDL's have been replaced by new ones, thus getting new excitement into the teaching. Wherever possible the new problems are slanted toward the applications of the material to daily life. We also have eliminated 4 experiments and added 3 new ones that do not require new equipment.

Most of the suggestions for changes originated in a two-day meeting with the following IPS teachers: Alexander Aliapoulios, Merle M. Bush, Harlan Downing, George E. Hall, Richard A. Hoag, Harold Pratt, and James A. Tulloch. We are greatly indebted to them for giving us the benefit of their long experience with the program. We wish also to express our appreciation to Dr. H. Graden Kirksey for his assistance in the development of the new experiments.

Susan McMahon, Editor for the Physical Science Group, and Edward M. Steele, Science Editor at Prentice-Hall, helped pull all the pieces together, and Myrna Goldblat prepared the new illustrations.

Acknowledgements to those who contributed to the writing and production of the First Edition of this book will be found in the Appendix.

August 1976

Uri Haber-Schaim

Jusdon B. Cross

Gerald L. Abegg

John H. Dodge

James A. Walter

Contents

1 Introduction 1

1.1 Experiment: Distillation of Wood 3

2 Volume and Mass 7

2.1 Volume 7
2.2 Reading Scales 11
2.3 Experiment: Measuring Volume by Displacement of Water 13
2.4 Shortcomings of Volume as a Measure of Matter 15
2.5 Mass 16
2.6 Experiment: The Equal-arm Balance 18
2.7 Experiment: Calibrating the Balance 19
2.8 Experiment: The Precision of the Balance 19
2.9 Experiment: The Mass of Dissolved Salt 21
2.10 Experiment: The Mass of Ice and Water 21
2.11 Experiment: The Mass of Mixed Solutions 21
2.12 Experiment: The Mass of Copper and Sulfur 22
2.13 Experiment: The Mass of Gas 23
2.14 The Conservation of Mass 23
2.15 Laws of Nature 25

3 Characteristic Properties 29

3.1 Properties of Substances and Properties of Objects 29
3.2 Experiment: Freezing and Melting 30
3.3 Experiment: Micro-melting Point 33
3.4 Experiment: Boiling Point 35
3.5 Density 36
3.6 Dividing and Multiplying Measured Numbers 37
3.7 Experiment: The Density of Solids 38
3.8 Experiment: The Density of Liquids 39
3.9 Experiment: The Density of a Gas 40
3.10 The Range of Density 43
3.11 Identifying Substances 44

4 Solubility 49

4.1 Experiment: Dissolving a Solid in Water 49
4.2 Concentration 50
4.3 Experiment: Comparing the Concentrations of Saturated Solutions 51
4.4 Experiment: The Effect of Temperature on Solubility 53
4.5 Wood Alcohol and Grain Alcohol 56
4.6 Experiment: Methanol as a Solvent 58
4.7 Sulfuric Acid 59
4.8 Experiment: Two Gases 59
4.9 Hydrogen 60
4.10 Carbon Dioxide 61
4.11 The Solubility of Gases 62
4.12 Experiment: The Solubility of Ammonia Gas 63
4.13 Other Solvents 64

5 The Separation of Substances 67

5.1 Experiment: Fractional Distillation 67
5.2 Petroleum 70
5.3 The Direct Separation of Solids from Liquids 73
5.4 Experiment: Separation of a Mixture of Solids 74
5.5 The Separation of a Mixture of Soluble Solids 76
5.6 Experiment: Fractional Crystallization 79
5.7 Experiment: Paper Chromatography 80
5.8 Mixtures of Gases: Nitrogen and Oxygen 80
5.9 Low Temperatures 82
5.10 Mixtures and Pure Substances 84

6 Compounds 90

6.1 Experiment: Decomposition of Sodium Chlorate 90
6.2 Experiment: Decomposition of Water 93
6.3 The Synthesis of Water 95
6.4 Experiment: Synthesis of Zinc Chloride 98
6.5 The Law of Constant Proportions 101
6.6 Experiment: A Reaction with Copper 102
6.7 Experiment: The Separation of Copper Oxide from a Mixture of Copper Oxide and Copper 104
6.8 Complete and Incomplete Reactions 104

7 Elements 107

7.1 Experiment: Zinc and Copper 107
7.2 Elements 108
7.3 Two Special Cases: Lime and Oxymuriatic Acid 111
7.4 Experiment: Flame Tests of Some Elements 113
7.5 Experiment: Spectra of Some Elements 113
7.6 Spectral Analysis 114
7.7 Radioactive Elements 118
7.8 Radioactive Decomposition 121
7.9 A Closer Look at Radioactivity 123

8 The Atomic Model of Matter 129

8.1 A Model 129
8.2 Experiment: A Black Box 131
8.3 The Atomic Model of Matter 132
8.4 "Experiment": Fasteners and Rings: Constant Composition 134
8.5 "Experiment": Some Other Compounds of Fs and R 136
8.6 A Prediction from the Atomic Model of Matter 137
8.7 Experiment: Two Compounds of Copper 138
8.8 The Law of Multiple Proportions 140
8.9 Molecules 144
8.10 Radioactive Elements and the Atomic Model 148

9 Molecular Motion 151

9.1 Molecular Motion and Diffusion 151
9.2 Density and Pressure of a Gas 154
9.3 A Prediction About the Relation Between Volume and Pressure of Gases 157
9.4 The Compressibility of Gases 160
9.5 Behavior of Gases at High Pressures 165
9.6 Temperature and Molecular Speed 167
9.7 Thermal Expansion of Gases 170
9.8 Thermal Expansion of Liquids 170
9.9 Experiment: Thermal Expansion of Solids 173
9.10 Molecular Motion in Liquids and Solids 175
9.11 Experiment: Growing Small Crystals 177

10 Sizes and Masses of Atoms and Molecules 183

10.1 The Thickness of a Thin Layer 183
10.2 Experiment: The Thickness of a Thin Sheet of Metal 185
10.3 Experiment: The Size and Mass of an Oleic Acid Molecule 186
10.4 The Mass of Helium Atoms 188
10.5 The Mass of Polonium Atoms 194
10.6 The Size of Polonium and Helium Atoms 197
10.7 Atomic Masses and Molecular Formulas 199

Epilogue 204

Appendix 205

Answers to Problems Marked with a Dagger (†)

Acknowledgments 207

Index 209

Third Edition Introductory Physical Science

Introduction

1

The scene at the center of the sun must be very monotonous. Everything must look exactly the same. There are no rocks, no trees, no rivers, no people. On earth, on the contrary, matter assumes an almost endless variety of forms. It is the variety of matter, the almost countless different substances that go together and behave in so many different ways, that makes the world interesting and enables living organisms to exist in it.

One can invent schemes that explain the endless variety of material things in terms of fewer, simpler things. Many different things could conceivably be made from the same units by putting them together in a variety of ways. For example, one can use bricks in many ways to make a wall or a house or a doorstop or a paved street. Over 2,000 years ago the Greek philosopher Democritus conceived of small units, which he called "atoms." But Democritus did not really know that there *were* atoms or how many different kinds there were. His ideas must have been important, because we still use the word atom; but the word in itself does not really explain anything. It did not help people to predict any properties of matter or to understand what kind of changes could or could not take place.

Modern chemistry and physics can give a much more meaningful account of the properties of matter. If this account is to have any meaning to you, we shall have to start at the beginning. We cannot just throw new words at you. Each step must be filled in with many experiments that you will perform. Then all the words and ideas will correspond to something real for you, and you will reach conclusions on your own.

You have been using this process—consciously or unconsciously—throughout your life. When you were a baby, signals from the outside world poured into you through your senses; but they had little meaning for you. You had to learn their meaning by physically reacting to your environment; that is, you learned by experimentation. You reached out for things, handled them when you could, threw them about. These early activities gave you a sense of size, of location, of weight. You began to associate what you saw with the properties of objects, so that you could begin to know something about objects through seeing them without having to handle them.

By the time you were two or three years old, signals from the outside world had acquired meaning for you: Particularly, you had learned to understand your parents' talk, and you had begun to talk yourself. After that, how rapidly your learning accelerated! By asking questions and understanding answers you could learn from other persons' experiences in addition to your own.

As time has gone on, you have continued to learn by experimenting and by asking questions and seeking answers. You cannot learn to ride a bicycle without getting on one. But it certainly helps to have the help of a friend who already knows how to ride.

So the study of science is really only a continuation of the activities that you have done so many times before. In this course you will be concerned chiefly with how to ask questions of nature and how to carry out experiments that will help you find some answers.

Therefore, let us first list some of the properties of matter that we can observe. Then we shall select a few of these properties and learn how to measure them and how to use them to distinguish between different materials. Finally we shall use them to put some real meaning into the word "atom."

Here is a list of questions that relate to some of the properties in which substances differ. Can you answer them?

A substance can be a solid, a liquid, or a gas. Are ice and water the same substance? What do you mean by substance?

A substance can have a high boiling point or a low boiling point. What substances have higher boiling points than water?

A substance can be more or less dense than another substance. What does this mean?

A substance can be an insulator or a conductor of electricity; a substance can be either brittle or malleable. Is there any connection between brittleness and electrical conductivity?

A substance can be strong or weak. Which is stronger, aluminum or copper? What do you mean by strong?

When you heat something, does its temperature always rise?

Some substances are more soluble than others in water or in alcohol. How can solubility be measured?

One of the best ways to find out how a thing works—and what it is made of—is to take it apart. Sometimes you can even test your understanding of it by trying to put it together again. But, of course, it matters how you take it apart. If you hit a watch with a hammer, it will come apart all right, but you will not learn much about how it works; and you certainly cannot put it back together. A great deal of modern experimental science is involved in learning how to take things apart in some instructive fashion. (There are, of course, some things, such as stars, that we have to learn about without being able to get at them.)

We shall start illustrating this way of learning about matter by taking something apart. We have chosen a common form of matter and one that yields many different substances when heated—namely, wood. We could, of course, take the wood apart with a knife or by burning it. The first method is too gentle and the second too rough for what we want to find out.

The things you do when you take wood apart and the questions that will come up will serve as a starting point.

EXPERIMENT DISTILLATION OF WOOD 1.1

What will happen if you heat some wood splints in a test tube without burning them? Try to predict what will happen before you read on.

Pack a Pyrex test tube with wood splints, and connect it to the apparatus as shown in Fig. 1.1. Heat the tube strongly with the two alcohol burners. What do you observe happening when it gets hot? Will the gas that comes out of the tube burn? Try lighting it.

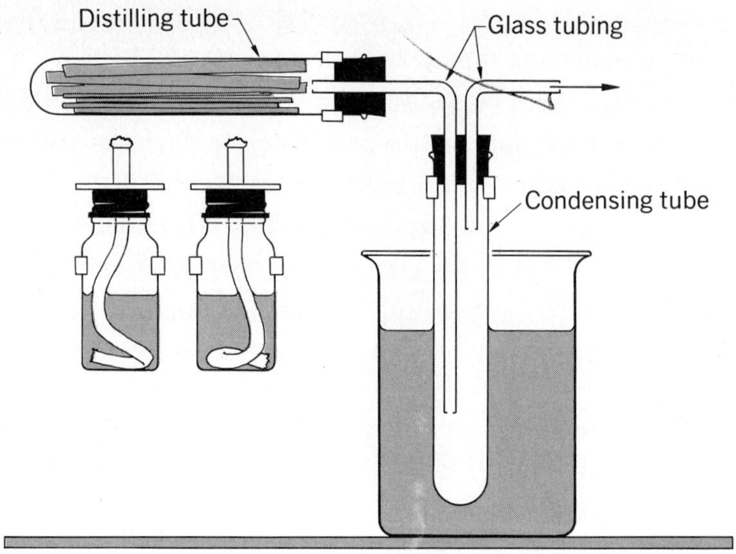

FIG. 1.1
Apparatus used in distilling wood. Lubricate the glass tubing with glycerin or water before pushing it into the stoppers. Hold the tubing with a towel as close to the stopper as possible while pushing it in. This will prevent the tubing from hurting you in case it breaks.

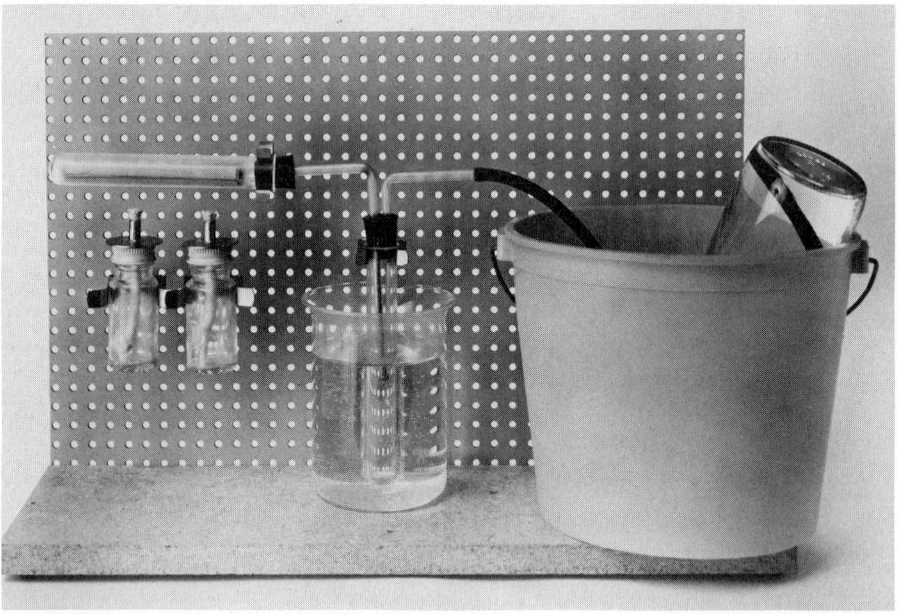

FIG. 1.2
Apparatus for collecting gas from the distillation of wood. The gas collects over water in the water-filled bottle on the right. The bucket should be about one-third full of water.

FIG. 1.3
Apparatus for obtaining more information about the distillation of wood. The burner-stand screen reduces the amount of heat reaching the test tube and keeps the liquid from boiling too rapidly.

Now attach a rubber tube to the apparatus, and collect the gas in a bottle by the displacement of water (Fig. 1.2). When the bottle is full of gas, disconnect the rubber tube, relight the gas coming from the test tube, and keep it burning as long as you can. As the gas may come out in puffs, you may have to relight it frequently. You probably will have to tilt or move the burners to heat all the wood.

Examine the liquid in the upright test tube. Is it one liquid or more than one? Disconnect the tube containing the condensed liquid, and put in a few pieces of broken porcelain. Connect the apparatus as shown in Fig. 1.3, and boil off about half of the liquid. The porcelain chips keep the liquid boiling evenly. What happens?

Is the liquid that condensed in the right-hand test tube the same as that in the left-hand one? Compare the two liquids. What happens if you mix them together?

After the test tube containing the wood has cooled, examine the remains of the wood splints. Try burning one. Does it leave any ash?

Could you predict, just by looking at and handling the wood, that all these gases and liquids could be obtained from it? Can you get the wood back by mixing all the stuff you have collected? Were these substances there all the time, or were they formed by heating? Have you any evidence for your answer? Why do you think you got different substances in the first place? How can you compare the amounts of the different solids, liquids, and gases that you got from the wood?

In order to get more definite answers to these questions, we shall do experiments, and these in turn will raise new questions. We shall start in the next chapter with the last question: How can we compare amounts of solids, liquids, and gases?

Volume and Mass

2

VOLUME 2.1

Suppose that you have some pennies stacked one on top of another in several piles and that you want to know how many pennies are in each pile. The obvious thing to do is to count them. If you had to count the pennies in many piles, you could speed up the counting in the following way: Make a scale like that shown in Fig. 2.1, marking it off in spaces equal to the thickness of one penny. You can then place this scale alongside each pile and read off the number of pennies.

FIG. 2.1
A scale for counting the number of pennies in a vertical pile. The distance between marks is the thickness of one penny.

If you want to measure the amount of copper in each pile of pennies, you first have to decide in which unit to measure the amount of copper. If you choose as the unit the amount of copper in one penny, then the amount in the whole pile is expressed by the same number as the number of pennies.

Suppose, now, that you want to find out how much copper there is in a rectangular solid bar of copper. You might think of making a box of the same size and shape as the copper bar and then counting the number of pennies needed to fill the box. This idea will not work because if you place pennies next to one another in a rectangular box, there will always be some empty space between them. A better way to measure the amount is to choose a new unit, such as the volume of a small cube. Suppose we had a box the same size and shape as the copper bar, and we could fill it with cubes of

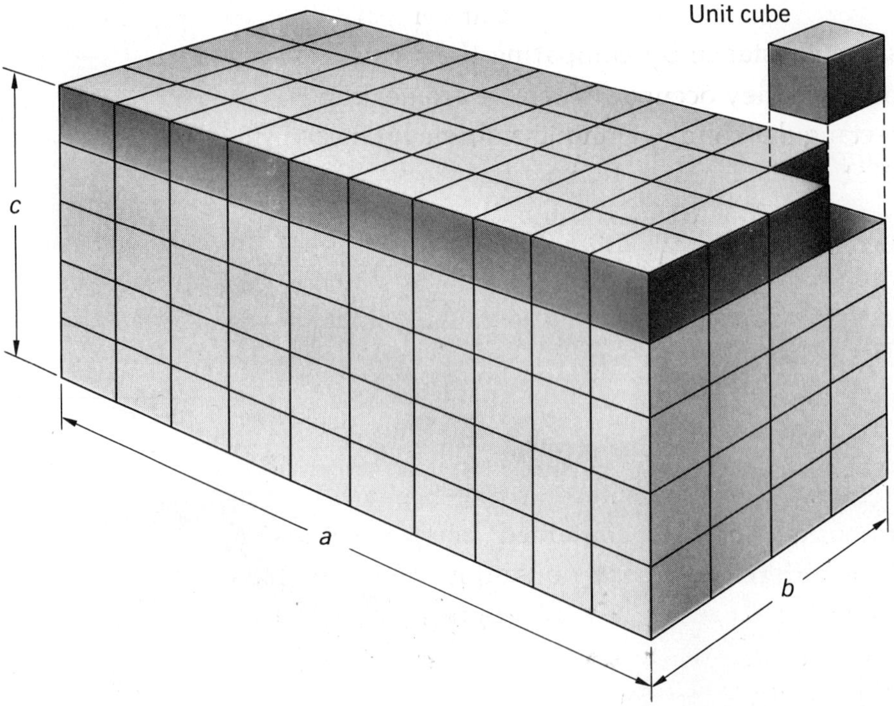

FIG. 2.2
A bar of copper 10 cubes long, 4 cubes wide, and 5 cubes high. One layer of the bar contains 10 rows of 4 cubes each, or 10×4 cubes. We see that there are 5 layers in the bar, each containing 10×4 cubes. The total number of unit cubes in the bar is therefore $10 \times 4 \times 5 = 200$ cubes. If the unit cube is 1 cm on a side, the volume of the bar is 200 cm^3 (cubic centimeters). For any rectangular solid, therefore, the volume is the product of the three dimensions, $a \times b \times c$.

copper of a size that would fit without air spaces between them. We could simply count the number of cubes to find the amount. Of course, we do not have to count each cube. If a cubes fit along the length of the box, b along the width, and c along the height, then the total number of cubes in the box (and bar) is $a \times b \times c$. (See Fig. 2.2.) This is the amount of copper in the solid bar, expressed in units of cubes. As you probably know, this is also the *volume* of the bar, expressed in terms of the volume of the unit cube.

What we choose to be the length of each side of this unit cube is a matter of convenience. We shall choose a unit of length based on the meter (m), the international standard of the metric system. In this case, as in much of our other work in this course, we shall use the centimeter (cm). A centimeter is $^1\!/_{100}$ m. Our unit cube would then be the cubic centimeter (cm^3), a small cube 1 cm on an edge.

To sum up, then, we can compare different amounts of the same substance by comparing their volumes—that is, the amounts of space they occupy. When we are dealing with a rectangular solid, we find its volume by measuring its three sides and taking the product of these numbers. We can also calculate the volumes of solids of other regular shapes from measurements of their dimensions, but this requires further knowledge of geometry.

The use of volume to compare amounts of substances is particularly convenient when we deal with liquids, because liquids take the shape of their containers. If we wish to compare two amounts of water contained in two bottles of very different shapes, we simply pour the contents of each separately into a graduated cylinder

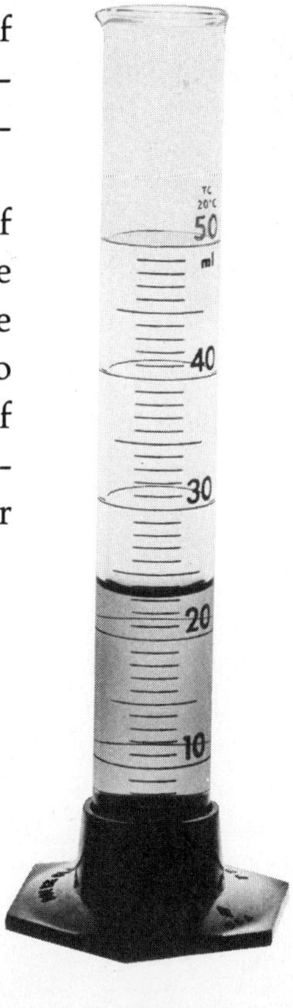

FIG. 2.3
A graduated cylinder marked off in units of volume. The cubic-centimeter marks could be made by filling the cylinder with liquid from a small cubic container, 1 cm on a side, and making a mark at the liquid level each time a containerful of the liquid is poured in. Many graduated cylinders are marked off in milliliters (ml). A milliliter is the same as a cubic centimeter.

that has already been marked off with the desired units; we then read off the volumes (Fig. 2.3). This way of measuring volume is very much like counting pennies all stacked up in a pile.

We can use the property of a liquid to take any shape when we want to find the volume of a solid of irregular shape, such as a small stone. After pouring some water into a graduate and reading its volume, we can submerge the stone in the water and read the combined volume of the water and the stone. The difference between the two readings is the volume of the stone.

1† A student has a large number of cubes that measure 1 cm along an edge. If you find it helpful, use a drawing or a set of cubes to answer the following questions.

a) How many cubes will be needed to build a cube that is 2 cm along an edge?

b) How many cubes will be needed to build a cube that is 3 cm along an edge?

c) Express, in cubic centimeters, the volumes of the cubes built in (a) and (b).

2 A student can carry two filled grocery bags. What is the approximate volume of each of these bags? How many cereal boxes can fit into each bag? What assumptions do you have to make to fit the boxes in the bag?

3 Figure A shows a graduate, used for measuring the volume of liquids, that is shaped like a cone. Why are the divisions not equally spaced?

FIG. A For prob. 3.

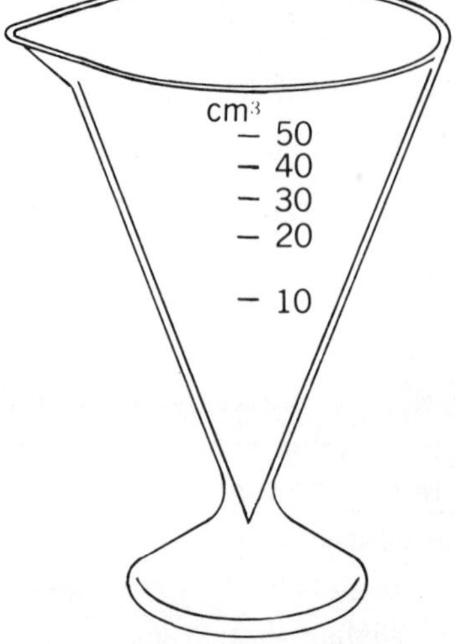

READING SCALES 2.2

To measure length with a ruler, volume with a graduated cylinder, and temperature with a thermometer, you must be able to read a scale. Therefore, learning how to get all the information a scale can provide is a useful skill to have.

We shall begin with reading a metric ruler (Fig. 2.4). The smallest divisions on such a ruler are 0.1 cm (1 mm) apart. This is a small distance indeed. Nevertheless, when the object you wish to measure has sharp edges, you can see whether the edge falls on one of the lines.

In Fig. 2.5 the edge falls between two lines. It is clear that the length is between 4.8 cm and 4.9 cm. To gain more information estimate the position of the edge. If you cannot tell whether the edge is closer to one line or the other, it is best to report the reading as 4.85 cm, or 48.5 mm.

If the edge is closer to the left line, report the reading as 4.82 cm or 4.83 cm. Either way you will not be off by more than ±0.02 cm. (The notation ± means "plus or minus.") Similarly, if you decide that the edge is closer to the right line, report the reading as 4.87 cm or 4.88 cm. Again you will not be off by more than ±0.02 cm. Had you read the scale as 4.8 cm or 4.9 cm, you might have been off by as much as 0.05 cm.

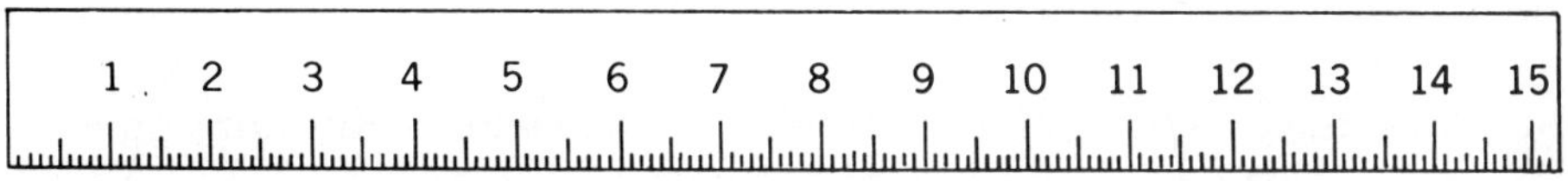

FIG. 2.4
A metric ruler. The numbered divisions are centimeters; the small divisions are $\frac{1}{10}$ cm, or millimeters.

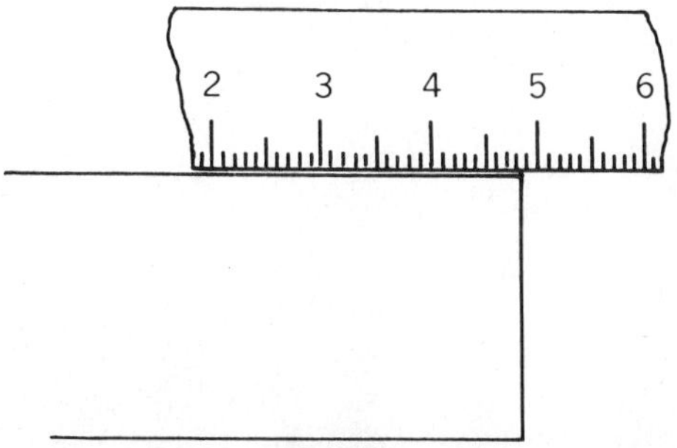

FIG. 2.5
Reading the position of the edge of an object. Here the edge falls between two of the millimeter marks.

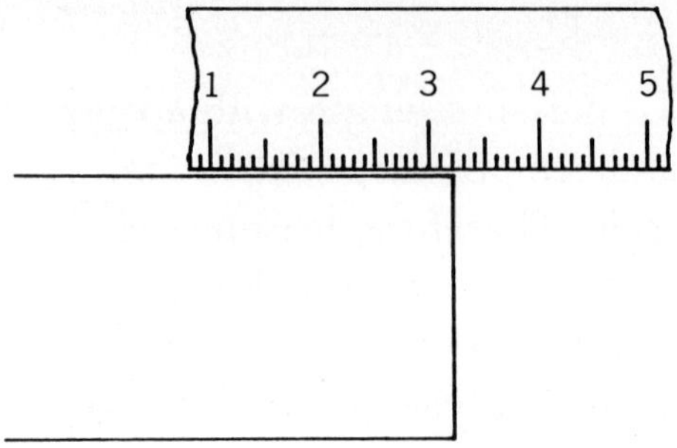

FIG. 2.6
Reading the position of the edge of an object. Here the edge falls on one of the millimeter marks.

Suppose, as far as you can tell, that the edge falls on a line (Fig. 2.6). Then you should report the reading as 3.20 cm. This will indicate that the reading is closer to 3.20 cm than to either 3.22 cm or 3.18 cm. Here the "0" gives us information that would be lost had you only written 3.2 cm.

---

4 The scale in Fig. B is in centimeters. Estimate the positions of arrows *a* and *b* to the nearest 0.1 cm. Can you estimate their positions to 0.01 cm? To 0.001 cm?

FIG. B For prob. 4.

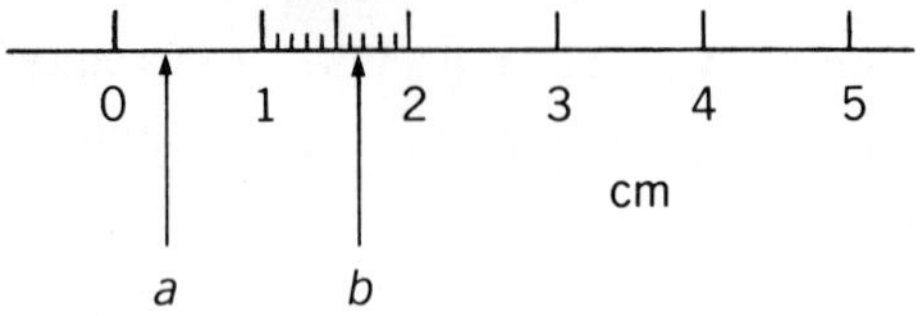

5† What fraction of a cubic centimeter do the smallest divisions on each of the graduated cylinders in Fig. C represent?

FIG. C For prob. 5.

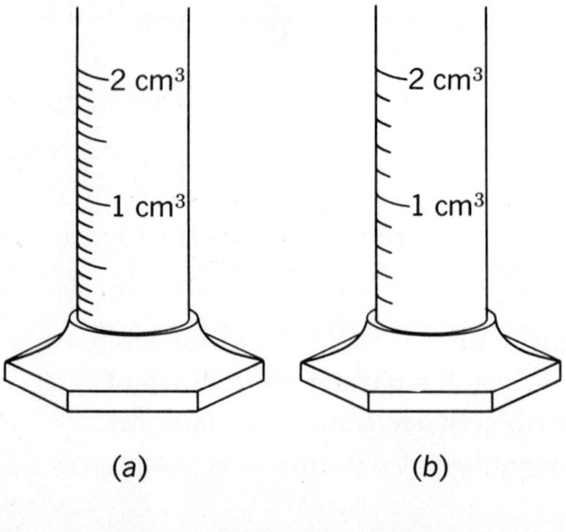

FIG. D For prob. 6.

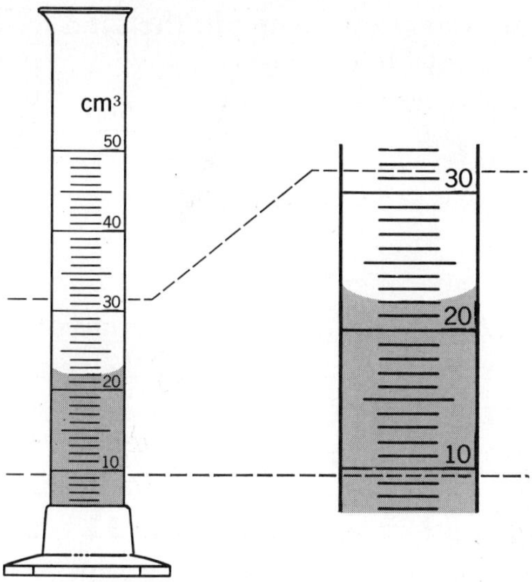

6 **A close look at Fig. D shows that the top of the liquid contained in the graduated cylinder is not flat but curved. How do you decide how much water is in the cylinder?**

7 **Three students reported the length of a pencil to be 12 cm, 12.0 cm, and 12.00 cm. Do all three readings contain the same information?**

8 **What advantage is there to making graduated cylinders narrow and tall rather than short and wide?**

EXPERIMENT MEASURING VOLUME BY DISPLACEMENT OF WATER 2.3

A granular solid like sand, although it does not flow so well as a liquid, can be measured by the same method. Suppose we have some sand in a cup. We can find how much space it takes up in the cup by simply pouring it into a graduated cylinder. But does the mark it comes to on the scale of the cylinder really show the volume of the sand? What about the air spaces between the loosely packed grains? What the graduated cylinder really measures is the combined volume of the sand plus the air spaces. However, we can do a simple experiment to find the volume of the sand alone.

Pour some sand into a dry graduated cylinder until it is about two-thirds full. What is the volume reading on the scale?

Now pour the sand into a beaker, and pour water into the graduated cylinder until it is about one-third full. Record the volume of the water, and then add the sand to the water. What is the volume of sand plus water?

From the last two readings calculate the volume of the sand alone. What is the volume of the air space in the sand? What fraction of the dry sand is just air space?

The experiment you have just done shows that we must be careful when we talk about the volume of a sample of a dry substance like sand. We must say how the volume was measured. If we have a bag of dry sand and want to know how many quart bottles it will fill, we need to know its volume dry. But if we want to know the volume of sand alone, not sand plus air space, then we must do an experiment like the one you have just done. We must measure the volume by liquid displacement.

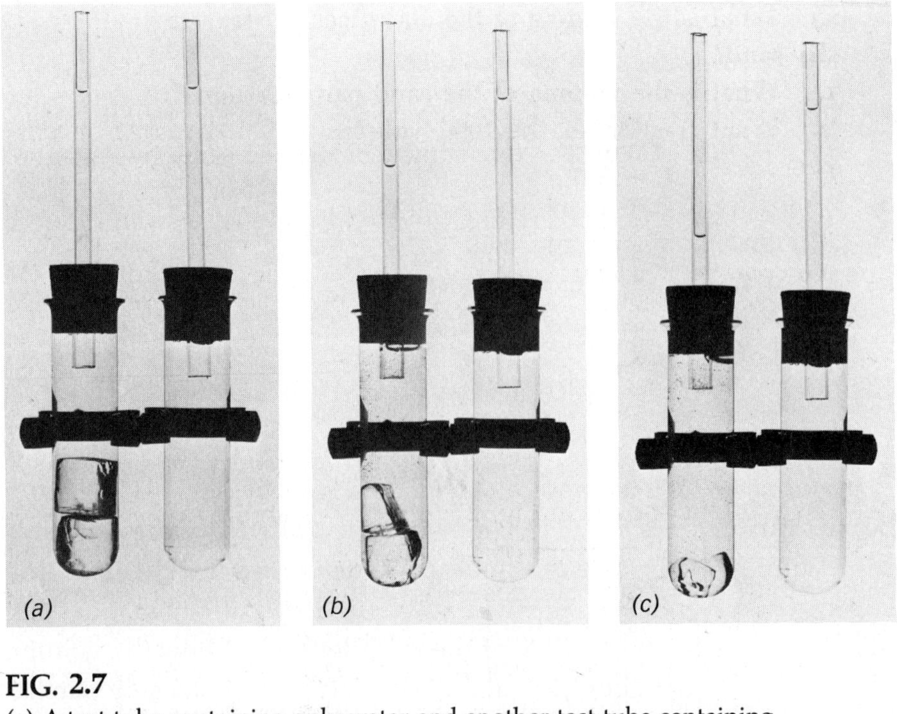

FIG. 2.7
(*a*) A test tube containing only water and another test tube containing water to which two large pieces of rock salt have just been added. (*b*) The same test tubes 30 min (minutes) later, after the salt has begun to dissolve. Notice the decrease in the total volume of the rock salt and water, as shown by the water level in the narrow glass tube. The test tube containing the salt was shaken several times to speed up the dissolving. (*c*) The test tubes after another 30 min. The total volume continues to decrease as more salt dissolves.

Whenever we measure the volume of a solid by displacement of water, we make the assumption that the volumes of the solid alone and of the water alone add up to the volume of the solid and water together. This assumption may or may not be correct. This will depend on the kind of solid we have. For example, if you measure the volume of a few chunks of rock salt by the displacement of water, you will see that the total volume of rock salt and water becomes less as the salt dissolves. (See Fig. 2.7.)

9† A volume of 50 cm^3 of dry sand is added to 30 cm^3 of water for a total volume of 60 cm^3.

a) What is the volume of water that does *not* go into air spaces between the sand particles?

b) What is the volume of water that does fill air spaces between the sand particles?

c) What is the volume of the air spaces between the particles in dry sand?

d) What is the volume of the sand particles alone?

e) What fraction of the total volume of the dry sand is sand particles?

10 How would you measure the volume of granulated sugar?

11 How would you measure the volume of a cork stopper?

SHORTCOMINGS OF VOLUME AS A MEASURE OF MATTER 2.4

The experiment shown in Fig. 2.7 strongly suggests that volume is not always a good measure of the amount of a substance. Here are some other difficulties: Suppose you wanted to find the amount of gas given off in your distillation of wood. You could measure the volume of gas you produced by filling bottles until no more gas was left. You could use one bottle as a unit of volume and express the volume as so many bottles of gas. Or you could collect the gas in inverted graduated cylinders instead of bottles and express the volume in cubic centimeters.

But if you have ever pumped up a bicycle tire, you know that a gas is very compressible. You know that, as you push more and more gas into the tire, its volume remains almost unchanged. Does this mean that the amount of gas in the tire remains almost unchanged, too? If you compressed the gas from the distillation of wood into a smaller volume, would there be less of it?

Finally, can we really use volume to compare the amounts of different substances, some of which may be solids, some liquids, and others gases? Consider again the distillation of wood. Does measuring the volume of the wood splints, the ashes, the liquids, and the gas really tell us how much of each of these substances we have?

2.5 MASS

The limitations of volume as a measure for the amount of matter must have been known to people many centuries ago because they developed a method for measuring the amounts of different substances quite independently of their volumes. From an Egyptian tomb several thousand years old, archaeologists have recovered a little balance arm of carved stone, with its carefully made stone masses (Fig. 2.8). It was almost surely used, in the very dawn of history, for the careful measurement of gold dust. Goldsmiths knew even then that the balance was the best way to determine the amount of solid gold they could cast from any heap of dust or from any pile of irregularly shaped nuggets.

The balance would be hung by the upper loop so that the horizontal bar was divided exactly into two arms of equal length. With no objects suspended from either arm, the balance bar would then hang horizontally. When an object was hung from the loop on the end of one arm, it could be balanced by hanging some other pieces of matter from the end of the other arm.

In using the balance, people soon learned, no doubt, that the bar would remain horizontal even though there were drastic changes in the shapes of the objects being balanced. Dividing a chunk of iron into a number of pieces or filing it into a pile of small grains does not affect the balance. A balance responds to something quite independent of the form of the object. What it responds to we call "mass."

Suppose a piece of gold balances a piece of wood, and the piece of wood balances a piece of brass. Then we say that the masses of all three are equal. If something else balances the piece of brass, it also balances the wood and the gold and therefore has the same mass. The equal-arm balance gives us a way of comparing masses

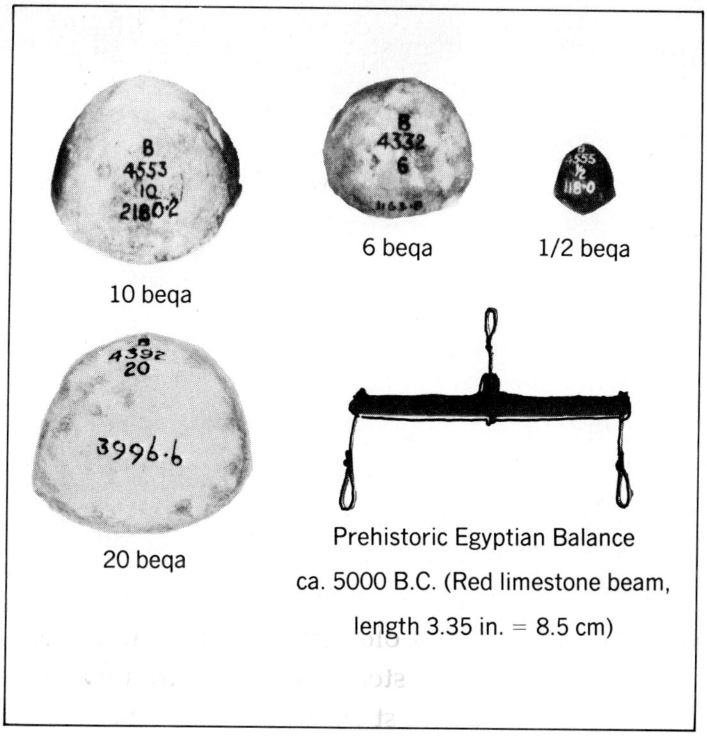

FIG. 2.8
This balance, the earliest one known, comes from a prehistoric grave at Naqada, Egypt. It may be 7,000 years old. The arm and masses are made of limestone. Other limestone masses of different numbers of *beqa* were also found in these prehistoric graves. Is there any reason why you should not use the *beqa* as your unit of mass? (*Courtesy of Science Museum, London.*)

of objects of any kind, regardless of their shape, form, or color or what substance they are made of.

To record masses we shall need some standard masses with which various other pieces of matter can be compared. This standard mass is arbitrary—any mass, even the ancient Egyptian *beqa*, can be chosen—but people must agree on it. In our work we shall use the gram (*g*), the fundamental unit of mass in the metric system. The international standard of mass in the metric system is a carefully made cylinder of platinum kept at Sèvres, near Paris, France, that has a mass of 1 kg (kilogram), or 1,000 g. All other kilogram masses are compared, directly or indirectly, with the standard whenever high precision is required. If we were to place a mass of 1 kg on the grocer's scale, the scale would read 2.2 pounds.

12 **When you buy things at the store, are they measured more often by volume or by mass? Give some examples.**

13 **What is your mass in kg?**

2.6 EXPERIMENT THE EQUAL-ARM BALANCE

In later experiments you will often use an equal-arm balance. The purpose of this experiment is to make you familiar with it and to allow you to develop the necessary skill in using it (Fig. 2.9).

Make sure that the pans swing freely and that the vertical pointer in the center does not rub against the support. The pointer of the balance should swing very nearly the same distance on each side of the center of the scale when there is nothing on either pan. In order to adjust the balance so that it swings in this manner, first make sure that the pointed metal rider on the right arm is as near to the center of the balance as possible. Then move the rider on the left arm until the long pointer on the center of the balance swings the same distance on each side of the center of the scale on the bottom of the balance.

FIG. 2.9
An equal-arm laboratory balance like the one you will use in your experiments. The object to be massed is placed on the pan at the left, and the standard gram masses are placed on the one at the right. The tip of the pointer hangs vertically down over the scale in the middle of the base.

Your balance comes with a set of masses, the smallest of which is 100 mg (1 mg, a milligram, is equal to 0.001 g. Thus 100 mg = 0.100 g). Now that your balance is adjusted, use a set of gram masses to mass several objects whose masses are between 1 and 20 grams. (We shall abbreviate "to find the mass of" by the verb "to mass.")

Exchange objects with your classmates, and compare the measurements you get with theirs.

EXPERIMENT CALIBRATING THE BALANCE 2.7

Look carefully at several pennies. Do you think they all have the same mass? Would you expect them to differ a little in mass? Now measure the masses of the pennies on your balance. Record the mass of each penny in a table in your notebook, and be careful to keep track of which penny is which.

You have massed the pennies only to the nearest 0.1 g. How can they be massed to less than 0.1 g to see if there are tiny differences in their masses, smaller than 0.1 g? By using the rider on the right arm of the balance, you can measure masses to less than 0.1 g. Move the rider until it balances a 0.1-g mass placed on the left-hand pan, and mark its position on the arm. Now make pencil marks on the arm, dividing into 10 equal spaces the distance between the 0 g and the 0.1-g position of the rider. Each mark represents an interval of 0.01 g on this rider scale. How can you check to see if this is true? If your balance has already been calibrated (that is, if there already is a scale marked on it), check to see if it is accurate.

EXPERIMENT THE PRECISION OF THE BALANCE 2.8

Now that you have calibrated your balance, again measure the mass of each penny. How do their masses compare? How much more precise is the balance when you use a rider than it was without a rider?

Since the space between the 0.01-g marks could also be divided into 10 equal spaces, does this mean that the balance masses accurately to 0.001 g (or 1 mg)?

To find out, mass separately to the nearest 0.001 g a light object and a heavy object. Make these massings several times, alternating light and heavy so that the balance must be readjusted for each massing. You do not have to wait for the balance to come to rest. It is necessary only that the pointer swing equal distances to the right and left of the center. What do you conclude?

14† Karen Jones massed an object three different times, using the same balance and gram masses. She obtained the following values: 18.324 g, 18.308 g, and 18.342 g. How could she best report the mass of the object?

15 A student masses an object on his balance. By mistake he places the object in the pan on the same side as the rider. He balances the object by means of 4.500 g in the opposite pan and by setting the rider to 0.060 g. What is the mass of the object?

16 Five students in turn used the same balance to measure the mass of a small dish; none knew what results the others obtained. The masses they found were

Student	Mass (g)
1	3.752
2	3.755
3	3.752
4	3.756
5	3.760

a) Can you tell whether any student made an incorrect measurement?

b) Do you think there is anything wrong with the balance?

c) What do you think would be the best way to report the mass of the dish?

d) How precise do you think the measurements were?

17 An equal-arm balance good to 0.01 g is used to mass two objects. If their masses are measured as 0.10 g and 4.00 g, what is the expected percentage error of each measurement?

18 You were told that your balance is an equal-arm balance, but suppose that the right arm is longer than the left arm. Would the object that is massed on your balance appear to have a mass that is greater than, less than, or the same as its true mass?

EXPERIMENT THE MASS OF DISSOLVED SALT 2.9

In Sec. 2.3 you learned that as salt dissolves in water the combined volume of salt plus water decreases. This leads us to ask whether the mass also decreases when salt is dissolved in water.

Pour about 2 g of salt into the cap of a small plastic bottle, and put it carefully aside. Pour water into the plastic bottle until it is about two-thirds full, and find the total mass of the bottle, water, cap, and salt when all are on the balance together but the salt and water are not mixed.

Pour the salt into the bottle, and put the cap on. What is the mass of the capped bottle of salt and water? Shake the bottle occasionally to speed up the dissolving of the salt.

Taking into consideration the precision of the balance, what do you conclude about the mass of salt and water as the salt dissolves?

19† In Expt. 2.9, how could you recover the dissolved salt? How do you think its mass would compare with the mass of dry salt you started with?

20† If the change in mass of the salt and water solutions was −0.0001 g in Expt. 2.9, would you have observed this change using your balance?

EXPERIMENT THE MASS OF ICE AND WATER 2.10

Here is another process where there is a volume change. When ice melts, it contracts—its volume decreases. Does its mass also change?

Mass a small container with its cover; then put in an ice cube, and mass again. What is the mass of the ice? After all the ice has melted (if the container is not transparent, you can tell by shaking it), mass again. Do you notice any condensation of water on the outside of the container? If so, what should you do about it? What do you conclude about change in mass when ice melts?

EXPERIMENT THE MASS OF MIXED SOLUTIONS 2.11

In the two experiments you have just done, a solid was either dissolved or melted. Now let us ask what happens to the mass when a solid is formed by mixing two liquids.

Your teacher will provide you with samples of two solutions. Find the total mass of the bottles of solution and their caps. Now pour one solution into the same bottle with the other, and cap both bottles. Again find the total mass of both bottles. Did the mass change as a result of the mixing?

2.12 EXPERIMENT THE MASS OF COPPER AND SULFUR

The changes you have examined so far were quite mild. A more drastic change in matter takes place when sulfur and copper are heated together. Does the total mass change when these substances are heated together?

Put about 2 g of granular copper [**CAUTION:** Do *not* use copper powder or copper dust.] and about 1 g of sulfur in a test tube, and close the end with a piece of rubber sheet held in place by a rubber band. Record the total mass of the closed tube. Heat the mixture gently until it begins to glow; then remove the flame immediately. (Let the test tube cool before you touch it.) Has the total mass of copper and sulfur changed?

Describe the appearance of the material in the test tube. Do you think the substance in the bottom is sulfur, copper, or a new substance?

21† The following data were obtained in an experiment in which copper and sulfur were reacted.

	Mass (g)
Tube and cover	20.484
Tube, cover, copper, and sulfur before reaction	23.440
Tube, cover, copper, and products after reaction	23.386

a) What is the mass of the substances before the reaction?
b) What is the apparent change in mass of the reacting substances?
c) What is the apparent percentage change in mass of the reacting substances?

22 **A test tube having 4.00 g of iron and 2.40 g of sulfur was heated in a manner similar to that of the copper-and-sulfur experiment. The total mass of the tube and contents measured on the balance before**

the heating was 36.50 g. After the heating, its mass was measured again. The mass of the tube and contents was 36.48 g.
a) Are you inclined to think it reasonable that mass was conserved during this experiment?
b) What additional steps would you take to strengthen your inclination?

EXPERIMENT THE MASS OF GAS 2.13

In this experiment a solid and a liquid produce a gas. Is there a change in mass?

First mass together a small *thick-walled* glass bottle one-third full of water, its cap, and one-eighth of an Alka-Seltzer tablet. Then place the piece of tablet in the bottle, immediately screw the cap on very tightly and place it back on the balance. Does what happens inside the bottle affect the mass of the bottle and its contents?

Slowly loosen the cap. Can you hear gas escaping?

Again mass the cap, the bottle, and its contents. What do you conclude?

THE CONSERVATION OF MASS 2.14

What have the last five experiments shown? If you have worked carefully, you have shown that all the changes in mass that you observed were within the experimental error of your equipment. Therefore your results agree with the conclusion that there was no change in mass that you could measure. From these experiments alone, you cannot predict with certainty that there will be no change in mass under other circumstances. For example, if we use larger amounts of matter in our experiments and use a balance of higher accuracy, we might measure a change greater than the experimental error. Then we would conclude that mass really does not remain the same. Furthermore, although we checked five rather different kinds of change, there is an endless variety of other reactions we could have tried, even more violent than the reaction of copper and sulfur.

What would happen, for example, if we set off a small explosion inside a heavy steel case, making sure no mass escapes? The experiments you have done give no direct answer to this question.

But we can make the guess that the results of these five experiments can be generalized in the following way: In all changes mass is exactly conserved, provided nothing is added (like the water that condensed on the outside of the closed container in the experiment with ice and water) or allowed to escape (like the gas in the last experiment). This generalization is known as the law of conservation of mass. It has been checked to one part in a billion* for a large variety of changes. That is, experiments have been done in which a change in mass of one billionth of the total mass would have been observed if it had occurred. Still, all this vast amount of evidence in favor of the law of conservation of mass does not prove that it will hold forever under all conditions. Surely, if someone claimed that he or she had done an experiment in which as much as one-millionth of the mass disappeared or was created, we should treat the results with great suspicion. First of all we should make many checks to determine whether there had been a leak of some sort in the apparatus from which, say, gas could escape. The chances are that we should find such a leak. On the other hand, if an experiment were done where a change in mass of one part in 100 billion was reported, we might have to conclude after a thorough examination of the experiment that the law of conservation of mass has its limitations, that it holds to one part in a billion but not to one part in 100 billion (10^{11}).

We have seen in this chapter that volume is very often a convenient way of measuring the quantity of matter. But we have also found out that, when matter changes form (when ice melts, salt dissolves, etc.), there is often an easily measurable change in volume but no observable change in mass: mass is conserved. It is the conservation of mass that makes mass such a useful measure of matter.

---

23 a) Express the following numbers in powers of 10.

100 10,000 100,000,000

b) Write the following numbers without using exponents.

10^5 10^6 10^9

*A billion is 1,000,000,000. Such a number is clumsy to write. Most of the zeros can be dispensed with by writing the number as 10^9 and reading it "ten to the ninth." The 9 is called an "exponent" and tells how many times we multiply 1 by 10 to get the number. For example, $1 \times 10 \times 10 = 10^2$, $1 \times 10 \times 10 \times 10 = 10^3$, etc. We shall use this way of expressing numbers, called "powers-of-10 notation," whenever it is convenient.

24 a) Express the following numbers in powers of 10.
1,000 5,280 93,000 690,000
b) Write the following numbers without using powers of 10.
5.0×10^3 10^7 1.07×10^2 4.95×10^4

LAWS OF NATURE 2.15

The law of conservation of mass is the first of several laws of nature that we shall study in this course. It is worthwhile to pause at this point and compare the laws of nature with the laws of our society. Laws of society are legislated; that is, they are agreed upon and then enforced. If evidence is presented that you have broken such a law, you are punished. The laws of society can also be changed or repealed.

Laws of nature are quite different. These are guessed generalizations based on experiments, often even crude experiments. If you do an experiment that appears to violate a law of nature, you are not punished. On the contrary, if you present convincing evidence that the law is not quite true, the law is changed to take into account the new experience. Only rarely does this amount to a complete repeal of the law; in most cases the change is a recognition of the limitation of the law.

25 There is an old saying, "Whatever goes up must come down." Does this express a law of nature? Why, or why not?

.0001 1000

For Home, Desk, and Lab

26 How would the volume of a piece of glass as measured by displacement of water compare with its volume as measured by displacement of burner fuel?

27 In determining the volume of a rectangular box, five cubes were found to fit exactly along one edge, and four cubes fit exactly along another edge. However, after six horizontal layers had been stacked in the box, a space at the top was left unfilled.
a) If the height of the space was half the length of a unit cube, what was the volume of the box?

FIG. E For prob. 28.

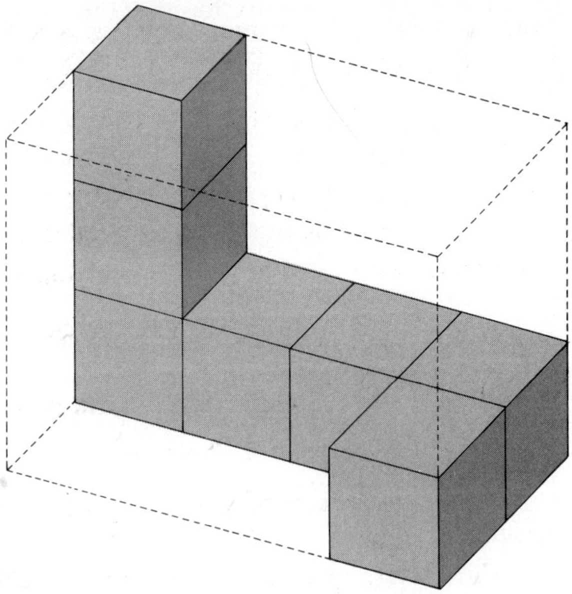

b) If the height of the space was 0.23 of the length of a unit cube, what was the volume of the box?

28 What is the total number of cubes that will fit in the space enclosed by the dashed lines in Fig. E? Is there more than one way to find an answer?

29 In an experiment in which the volume of dry sand is measured by the displacement of water, the sand was slightly wet to begin with. What effect would this have on the volume of air space that was calculated? On the percentage of the volume that was air space?

30 a) How would you measure the volume of a sponge?
b) What have you actually measured by your method?
c) Does this differ from your measurement of the volume of sand?

31 a) Completely fill two small bottles with water. Pour the water into a single larger vessel. Now refill the bottles with the same water. Are they both filled completely?
b) Now do the same thing again, but fill one bottle with water and the other with burner fuel. Compare the total volume of burner fuel and water before and after they were mixed together and poured back into the bottles. Is volume a good measure for the quantity of matter in this case?

32 Fuel oil usually is sold by the gallon, gas for cooking by the cubic foot, and coal by the ton. What are the advantages of selling the first two by volume and the last by mass?

33 In the following list of ingredients for a recipe, which are measured by volume, which by mass, and which by other means?

1½ pounds ground chuck	pinch of pepper
1 medium-size onion	3 drops Worcestershire Sauce
½ cup chopped green pepper	oregano to taste
4 slices day-old bread	3 tablespoons oil
1 teaspoon salt	1 1-pound can tomato sauce

34 a) What is the volume of an aluminum cube whose sides are 10 cm long?
b) What is the mass of the aluminum cube? (One cubic centimeter of aluminum has a mass of 2.7 g.)

35 One cubic centimeter of gold has a mass of 19 g.
a) What is the mass of a gold bar 1.0 cm × 2.0 cm × 25 cm?
b) How many of these bars could you carry?

36 Suppose that you took a balance home. When you were ready to use it, you found that you had forgotten a set of gram masses.
a) How could you make a set of uniform masses from materials likely to be found in your home?
b) How could you relate your unit of mass to a gram?

37 Figure 10.10 (p. 196) shows a sensitive equal-arm balance. Suggest reasons why it is enclosed in a case and why it is used with the sliding front cover closed.

38 You wanted to find the mass of water in a plastic bottle, and you took the following measurements using your equal-arm balance.

Mass of bottle and water	21.48 g
Mass of empty bottle	9.56 g
Mass of water	11.92 g

After you completed your measurements and calculations, you saw that you forgot to set the left-hand rider correctly; the beam was not level when the right-hand rider was on the zero mark and nothing was on the pans of the balance. Must you repeat the measurements to obtain the mass of the water?

39 a) You can compare your standard masses on the equal-arm balance after you have carefully adjusted it. How closely is the 50-gram mass equal to the sum of the 20-, 10-, 10-, 5-, 2-, 2-, and 1-gram masses? How closely is the 5-gram mass equal to the sum of the two 2-gram masses and the 1-gram mass? You can make other comparisons as well.

b) When you are finding the difference in mass between an empty container and the container filled with liquid, why should you try to use as nearly as possible the same particular masses for both measurements?

40 Suppose you lost the rider for your scale. Try to think of another method, not using a rider, by which you could measure hundredths of a gram.

41 Suppose you balance a piece of modeling clay on the balance. Then you reshape it. Will it still balance? If you shape it into a hollow sphere, will it still balance?

42 Suggest a reason for putting the lid on the small container that you used in studying the mass of ice and water.

43 In Expt. 2.10 would the mass of the container and its contents stay the same if you started with water and froze it? Try it.

44 Estimate in grams the mass of a wristwatch. Now find the mass of a nickel (5¢) on your balance. Estimate the mass of the watch again. Did you change your estimate? Does knowing the mass of a nickel help you to better estimate your own mass? Why?

45 Two astronauts on the moon use an equal-arm balance to find the mass of a specimen of moon rock; the specimen has a mass of 35.83 grams. When the astronauts return to earth and mass the specimen once again, will they find that the mass of the rock on earth is more than, equal to, or less than the 35.83 grams they measured on the moon?

Characteristic Properties

3

PROPERTIES OF SUBSTANCES AND PROPERTIES OF OBJECTS 3.1

How do we know when two substances are different? It is easy enough to distinguish between wood, iron, and rock or between water and milk; but there are other cases in which it is not so easy. Suppose that you are given two pieces of metal. Both look equally shiny and feel equally hard in your hand. Are they the same metal? Or think of two glasses containing liquids. Both liquids are transparent and have no smell. Are they the same or different?

To answer such questions we shall have to do things to substances that will reveal differences not directly apparent. Merely massing the two pieces of metal will not do; two objects can be made of different materials and yet have the same mass. Think, for example, of a 100-g steel cylinder and a 100-g brass cylinder of the kind used as masses on a balance. On the other hand, two objects can have different masses and be made of the same materials: for example, two hammers, both made of steel but one much larger and with a greater mass than the other. Mass is a property of an object; it is not a property of the substance of which the object is made.

To find out if two pieces of metal that look alike are made of the same substance, you may try to bend them. Again, one can be thick and hard to bend, and the other can be thin and easy to bend. Yet they can both be made of the same substance. On the other hand, you may find that two pieces of metal of different thicknesses but made of different substances bend with equal ease. Thus the ease of bending is also a property of the object and not of the substance.

If we want to find out whether two objects are made of the same substance or of different ones, we have to look for properties that are characteristic of a substance—that is, properties that do not depend on the amount of the substance or on the shape of the sample. In this chapter we shall concentrate on properties that show differences between substances. These we call characteristic properties.

1 In the following descriptions state which are properties of the substance and which are properties of the object.
a) A sharp, heavy, shiny, stainless-steel knife
b) A small chunk of black tar
c) A beautifully carved wooden chair

3.2 EXPERIMENT FREEZING AND MELTING

If you live in a part of the country where it snows in the winter, you know that a big pile of snow takes longer to melt than a small one. Does this mean that the big pile melts at a higher temperature? Let us see whether the temperature at which a sample of a substance melts or freezes is really a characteristic property of the substance. To do so, we shall measure the freezing temperatures of some substances by using samples of different mass. For convenience we shall use substances that freeze above room temperature.

Fill a test tube one-third to one-half full of moth flakes or nuggets and immerse it in a water bath. Heat the water until the solid in the test tube is completely melted. Insert a thermometer into the liquid. Make sure that the solid in the test tube is completely melted before removing the burner. For comparison it may be interesting also to measure the temperature of the water with a second thermometer. (See Fig. 3.1.) While the liquid cools, measure and

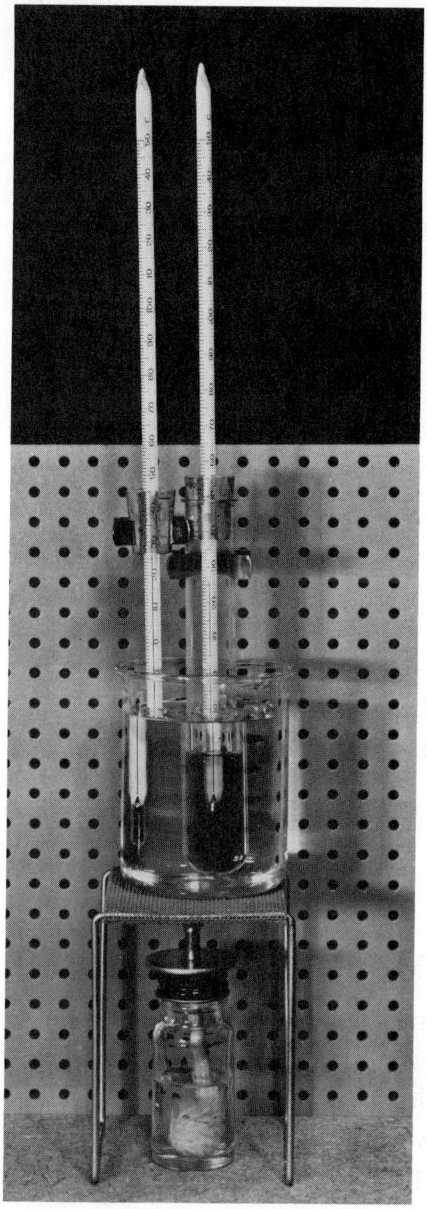

FIG. 3.1
Apparatus used to obtain data for the cooling curve of a liquid as it cools and freezes. The thermometer in the test tube measures the temperature of the liquid; the one in the beaker measures that of the water bath. The thermometers are calibrated in degrees Celsius (°C).

record both temperatures every half minute. (Stirring the water will ensure that the temperature will be the same throughout the water when you read it.)

In addition, record the temperature of the molten substance just as it begins to solidify. Continue to take data every half minute for about 5 min (minutes) after the substance has all solidified. On the same graph, plot both the temperature of the substance and the temperature of the water as a function of time, and compare your graph with those of your classmates. Do all the graphs have a flat section? Does the temperature of the flat section depend on the mass of cooling material? Do you think that all the samples used in the class were of the same material?

In this experiment you determined the freezing point of a substance by the plateau (flat section) in the cooling curve. With some substances the plateau is more easily recognizable than it is with others. Some cooling curves, however, may not have a flat section at all. For example, examine the cooling curve of candle wax shown in Fig. 3.2. The data for this curve were obtained in the same way as in your experiment. The fact that no part of the curve is flat means that candle wax has no freezing point; that is, there is no temperature at which it changes from liquid to hard solid without continuing to cool down during the process. Similarly, as you warm a piece of candle wax in your hand, it becomes softer and softer, but there is no temperature at which it changes from hard solid to liquid without continuing to warm up.

It is harder to measure the melting point of a substance than to measure the freezing point; since we cannot stir a solid, it is necessary to heat it very slowly and evenly. If, however, we do very

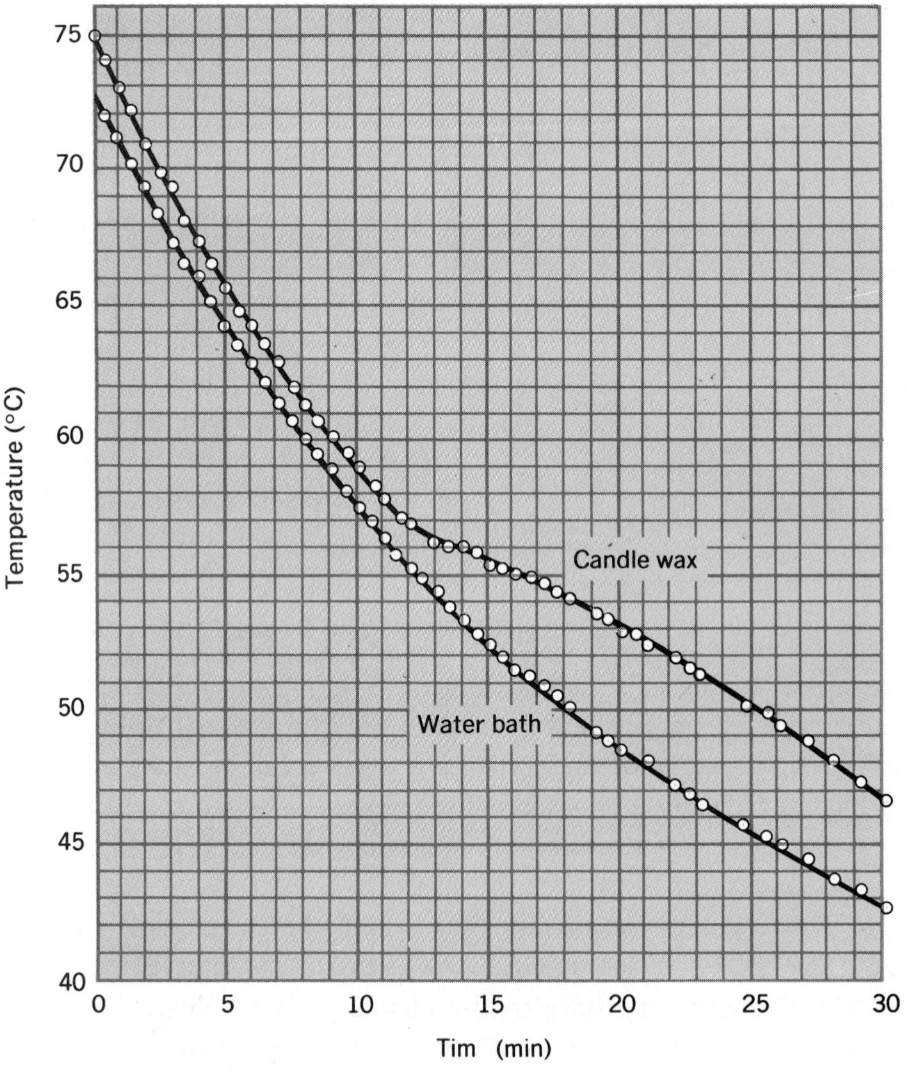

FIG. 3.2
The cooling curves of candle wax and of the water bath surrounding the test tube holding the wax. The absence of a flat section in the curve for the wax means that candle wax has no freezing point.

careful experiments to measure the melting point of a solid by heating it until it melts, we find that we get a curve with the flat portion at exactly the freezing temperature. A solid melts at the same temperature as its liquid freezes.

2† The graph in Fig. A represents data from an experiment on the cooling of paradichlorobenzene. During which time intervals is there (a) only liquid, (b) only solid, and (c) both liquid and solid?

FIG. A For prob. 2.

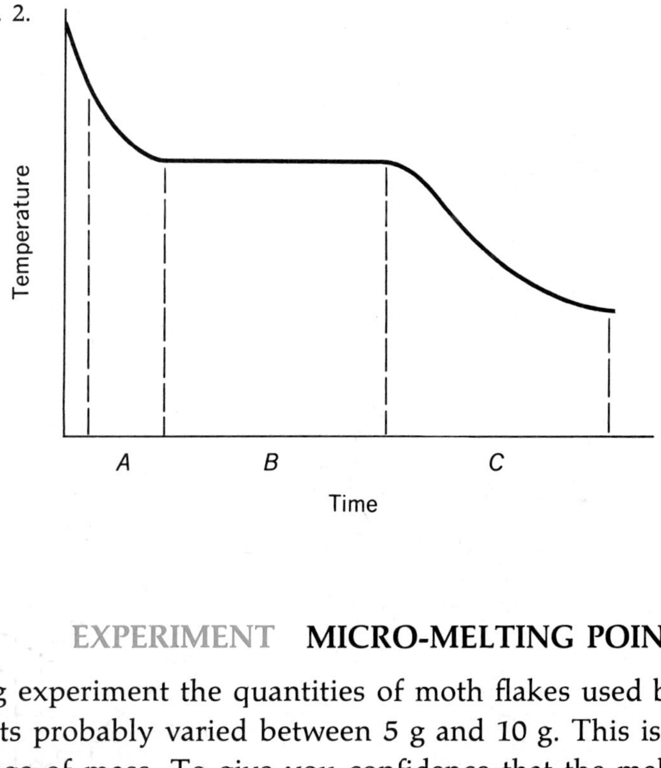

EXPERIMENT MICRO-MELTING POINT 3.3

In the preceding experiment the quantities of moth flakes used by different students probably varied between 5 g and 10 g. This is a rather small range of mass. To give you confidence that the melting point is really a characteristic property, it is worthwhile to repeat the experiment with a *much* smaller sample, with only a few tiny crystals.

To do this you need a very small tube closed at one end to hold the crystals. Prepare this tube in the following way: Heat the center of a capillary tube in the alcohol flame. When the center melts, pull the two pieces apart, break off the glass thread, and seal the end of each piece in the flame. Between your fingers crush a small crystal of the material you used in the preceding experiment. Scoop up two or three bits of the smaller pieces in the open end of one of the capillaries. By gently tapping the sealed end on the desk,

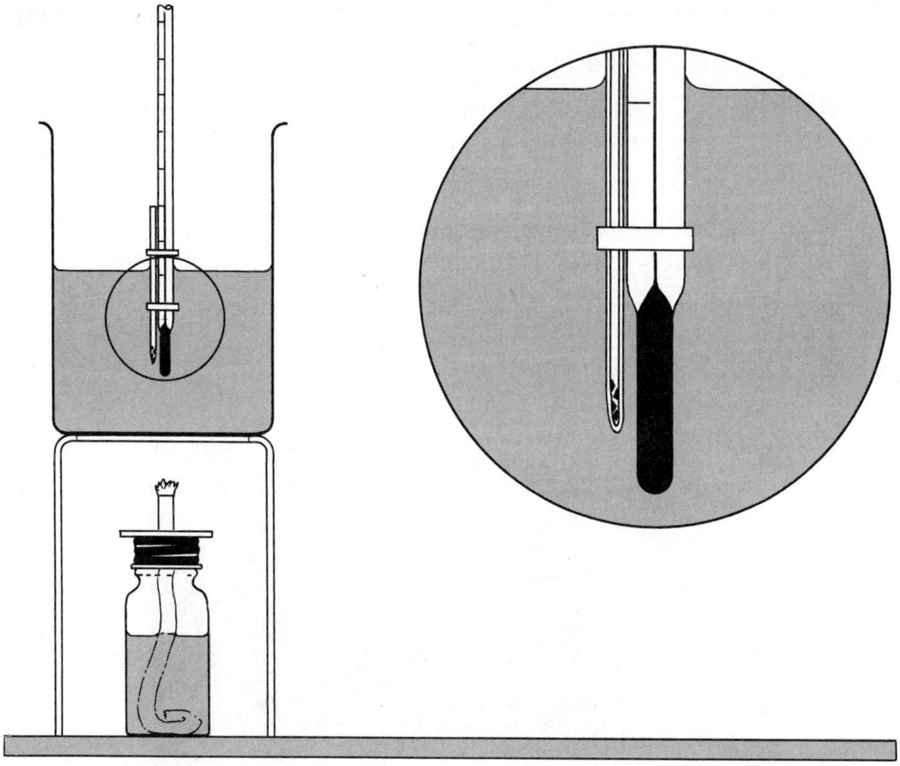

FIG. 3.3
The melting point of a few tiny crystals of a substance can be measured by supporting a small capillary tube containing the crystals next to the bulb of a thermometer in a water bath.

you can make the tiny crystals fall to the bottom. Fasten the tube to the side of a thermometer with two rubber bands so that the crystals are next to the thermometer bulb. (See Fig. 3.3.)

Support the thermometer in a beaker half-filled with water. Very slowly heat and stir the water while constantly observing the crystals. Read the thermometer the instant the crystals melt. At what temperature do the crystals melt? You may need to use the other half of the capillary tube to make a second and more careful determination.

How does the melting point compare with the freezing point you found when you cooled a large mass of the same substance? Estimate how many times larger the large mass was than one of the small crystals you melted. Does the melting point of a substance depend on the mass of the sample you use when you measure the melting point? Is it a characteristic property?

EXPERIMENT BOILING POINT 3.4

Everybody knows that it takes longer to get a full pot of coffee to boil than a half-filled one. Does this mean that the full pot gets hotter? To see what happens, heat about a fourth to a half of a test tube of a liquid in the apparatus shown in Fig. 3.4. To prevent uneven boiling, add a few small chips of porcelain to the liquid. The glass and the rubber tubes will prevent vapors from spreading into the room. The vapors will condense in the cooled test tube.

FIG. 3.4
A thermometer supported in the test tube measures the temperature of the liquid as it is heated to its boiling point. To prevent uneven boiling, a few small chips of porcelain are placed in the liquid.

Read the temperature of the liquid every half minute until about half the liquid has boiled away. Then plot a graph of the temperature of the liquid as a function of the time.

Compare your results with those of other students in your class. Do all the graphs look alike at the beginning? Do all the graphs have a flat section? Was the temperature the same in all test tubes once the liquid started boiling? Does the boiling point of a liquid depend on the amount of liquid? What does a difference in boiling point reveal? Is boiling point a characteristic property?

3.5 DENSITY

Suppose we cut up a piece of aluminum rod into sections of equal volume, say, 1 cm^3. We find that they all have the same mass when massed on a balance, no matter from what part of the rod they come. What if we take many 1-cm^3 samples from a bottle of water? We find that each cubic centimeter of water has the same mass. However, the mass of 1 cm^3 of water is different from the mass of 1 cm^3 of aluminum rod. That is to say, the mass of a unit volume of material is the same for all samples of the same substance but usually differs for different substances. The mass of a unit volume is, therefore, a characteristic property of a material. It can be used to distinguish one substance from another.

Rarely do we measure directly the mass of a unit volume of a sample of a substance. Usually we find that the volume of the sample is either larger or smaller than one unit volume. However, we can find the mass of a unit volume indirectly. We do this by measuring both the sample's mass and its volume. We then calculate the mass of one unit volume by dividing its mass by its volume. For example, consider a 30-g sample whose volume is 10 cm^3. The mass of 1 cm^3 of the material will be 30 g/10 = 3.0 g. Since each unit volume of a sample of a substance has the same mass, this indirect procedure will always give the same value for the mass of a unit volume as we would get by massing a sample whose volume is, in fact, one unit volume.

Because we generally find the mass of a unit volume by dividing mass by volume, we refer to it as mass *per* unit volume. (The word *per* means a division by the quantity that follows it. For example, the speed of a car is stated in miles *per* hour; that is, distance divided by time.) The mass per unit volume of a substance is called the *density* of the substance. Its units are g/cm^3 (grams per cubic centimeter). In the example mentioned (p.36) the density of the substance is $30\ g/10\ cm^3 = 3.0\ g/cm^3$. The two statements "the mass of 1 cm^3 of the substance is 3.0 g" and "the density of the substance is 3.0 g/cm^3" contain the same information. The second statement, however, is more concise.

---

3 A parking lot is filled with automobiles.

a) Does the number of wheels in the lot depend upon the number of automobiles?

b) Does the number of wheels per automobile depend upon the number of automobiles?

c) Is the number of wheels per automobile a characteristic property of automobiles that distinguishes them from other vehicles?

4 a) Draw a graph of the number of wheels in a parking lot as a function of the number of cars.

b) Draw a graph of the number of wheels per car as a function of the number of cars in the lot.

3.6 DIVIDING AND MULTIPLYING MEASURED NUMBERS

When you find the density of a substance from the measured mass and volume of a sample, you calculate with numbers of limited accuracy. For example, consider a pebble that has a mass of 12.36 g and a volume of 4.7 cm^3. Here the mass is given to four digits and the volume to two. If you calculate the quotient $12.36\ g/4.7\ cm^3$ on a calculator, you get the number 2.62978723. (You can, of course, do the division by hand to that many digits if you have the patience.) However, not even a calculator can produce numbers that are more accurate than the data used in the calculations.

For division and multiplication, it is good to remember a simple rule of thumb: The result should have as many digits as the measured number with the smallest number of digits. It is always advisable to calculate one additional digit, and then round off. In the example we just saw, the density is

$$\frac{12.36\ g}{4.7\ cm^3} = 2.6\ g/cm^3$$

The only significant digits are 2.6. The remaining digits in the display of the calculator are not significant and should be dropped.

If we calculate the volume of an object from its dimensions, the same rule applies. If the measured dimensions of a rectangular solid are 4.82 cm, 11.05 cm, and 1.28 cm, then the volume is correctly reported as

$$4.82 \text{ cm} \times 11.05 \text{ cm} \times 1.28 \text{ cm} = 68.2 \text{ cm}^3$$

(68.17 cm^3 is rounded off to 68.2 cm^3.) It should *not* be reported as 68.174080 cm^3.

How do we count the digits in a decimal measurement whose last digit is a zero, as in the measurement 4.20 cm? (Sec. 2.2 explained why the zero is written.) In these cases the zero is counted as a significant digit. However, in such measured numbers as 0.86 cm and 0.045 g, the zeros are not counted. Their only purpose is to locate the decimal point. Both these numbers are given to two significant digits.

---

5 **Calculate the following quotients to the proper number of digits.**

a) $\dfrac{125}{23.7}$ **c)** $\dfrac{0.065}{32.5}$

b) $\dfrac{20.5}{51.0}$ **d)** $\dfrac{1.23}{0.72}$

6 **Calculate the following products to the proper number of digits.**

a) 4.72×0.52

b) 6.3×10.08

c) $1.55 \times 2.61 \times 5.3$

d) $3.01 \times 5.00 \times 25.62$

3.7 EXPERIMENT THE DENSITY OF SOLIDS

Pick up two cubes that look alike and have the same volume. Can you decide by handling them whether they have the same or different masses?

Measure the masses of the cubes on your balance. Which has the greater density?

Now, just by handling them, compare the mass of each of the cubes with that of a third object of different volume. Can you decide in this way if the third object is made of the same substance or if it is made of a different substance?

Measure the dimensions of each of the three objects as accurately as you can. Calculate the volume of each and then their densities. Can you now decide whether the third object is made of the same substance as either of the cubes?

If you have an irregularly shaped object whose volume is difficult to determine from measurements of its dimensions, you can find its volume by the displacement of water, as described in Chap. 2.

Find the density of an irregularly shaped stone. Compare the density of your stone with the results of other students who have used pieces from the same rock. What possible reasons could you give for the different measured values of density?

7† What measurements and what calculations would you make to find the density of the wood in a rectangular block?

8 A student announced that she had made a sample of a new material that had a density of 0.85 g/cm^3. How large a sample had she made?

9† A block of magnesium whose volume is 10.0 cm^3 has a mass of 17.0 g. What is the density of magnesium?

10 Two cubes of the same size are made of iron and aluminum. How many times as heavy is the iron cube than the aluminum cube? (See Table 3.1, page 43.)

11† a) A 10.0-cm^3 block of silver has a mass of 105 g. What is the density of silver?
b) A 5.0-cm^3 block of rock salt has a mass of 10.7 g. What is the density of rock salt?
c) A sample of alcohol amounting to 0.50 cm^3 has a mass of 0.41 g. What is its density?

EXPERIMENT THE DENSITY OF LIQUIDS 3.8

Examine two samples of liquid. Smell them and shake them, but don't taste them. Can you tell whether they are the same or different? Perhaps by finding their densities you can answer the question. You can find the density of a liquid by massing it on a balance and measuring its volume with a graduated cylinder. The cylinder is too large to fit easily on the balance. You must mass the liquid in something smaller.

A small amount of liquid will stick to the inside of any container from which you pour it. Therefore, to be sure you mass the volume of the liquid you measure in the graduated cylinder, you must be careful of the order in which you make your measurements of mass and volume. Is it more accurate to mass the liquid in the small container before pouring it into the graduated cylinder or to determine its volume first?

Find the densities of the two liquids. Are they the same or different?

---

12 You are given two clear, colorless liquids. You measure the densities of these liquids to see whether they are the same substance or different ones.

a) What would you conclude if you found the densities to be 0.93 g/cm^3 and 0.79 g/cm^3?

b) What would you conclude if you found the density of each liquid to be 0.81 g/cm^3?

3.9 EXPERIMENT THE DENSITY OF A GAS

It is more difficult to measure the density of a gas than that of a liquid or a solid. Gases are hard to handle, and most of them cannot even be seen. In fact, early chemists neglected to take into account the mass of gases produced in experiments.

When we mix Alka-Seltzer tablets and water, a large volume of gas is produced. We can find the density of this gas by massing the tablets and the water before and after they are mixed and by collecting and measuring the volume of the gas. You will recall that in Expt. 2.13 you measured the mass of some of the same gas but you did not measure the volume.

Place two half tablets of Alka-Seltzer and a test tube containing about 10 cm^3 of water on the pan of your balance as shown in Fig. 3.5, and find the total mass of these objects.

Arrange the apparatus as shown in Fig. 3.6 so that you can collect the gas that will be evolved. Be sure the end of the rubber tube is at the top of the collecting bottle. When ready, drop the two

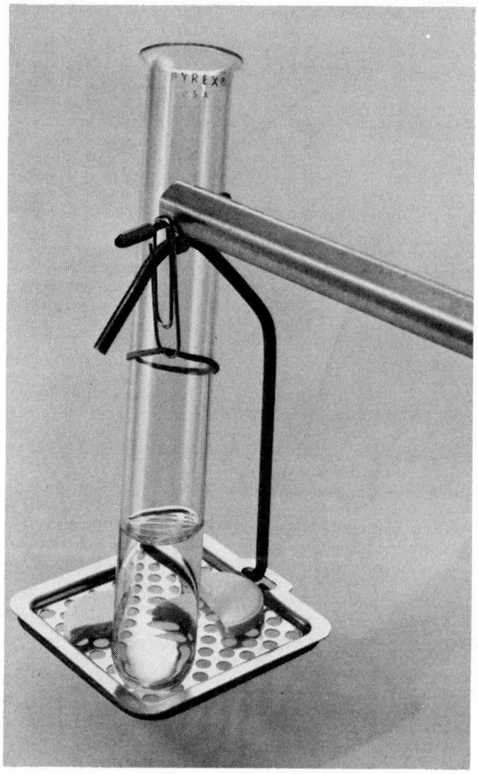

FIG. 3.5
To support a test tube containing water on the balance, you can use a paper clip and a rubber band as shown. Be sure that the paper clip does not rub against the arm of the balance.

FIG. 3.6
When the two half tablets are added to the test tube, the gas generated is collected by displacing water from the inverted bottle on the right.

half tablets into the water, quickly insert the delivery tube and stopper into the test tube, and collect the gas produced. Practically all the gas will be produced in the first 10 min of the reaction. Why is it important to hold your hand across the mouth of the bottle while removing it from the pan? Turn the bottle upright, and find the volume of the water displaced by the gas. How is this volume related to the volume of the gas?

Again mass the test tube and its contents. Why is the mass less than before? What is the density of the gas? What assumptions have you made in using this method?

13† A mixture of two white solids is placed in a test tube, and the mass of the tube and its contents is found to be 33.66 g. The tube is stoppered, and arrangements are made to collect any gas produced. When the tube is gently heated, a gas is given off, and its volume is found to be 470 cm^3. After the reaction, the mass of the test tube and its contents is found to be 33.16 g.

a) What is the mass of the gas collected?

b) What is the density of the gas collected?

14 Experiment 3.9 is repeated with a sample of a different solid. Here are the data obtained:

Mass of solid, test tube, and water before action . . . . . . 35.40 g
Mass of test tube and contents after action . . . . 34.87 g
Volume of gas collected . . . . 480 cm^3
Could this gas be the same as that produced in Expt. 3.9?

15 The volume of gas generated by treating 1.0 g of magnesium carbonate with 8.8 g of sulfuric acid is 200 cm^3. The remaining acid and solid have a mass of 9.4 g. What is the density of the gas evolved?

16 If the volume of gas in the preceding problem is compressed to 50 cm^3, what will the density of the gas now be? To what volume must the gas be compressed before it will reach a density of 1.0 g/cm^3, a typical density of a liquid?

17 The gas whose density you measured in the experiment of Sec. 3.9 dissolves slightly in water.

a) How does this affect the volume of the gas you collect?

b) How does this affect your determination of the density of the gas?

THE RANGE OF DENSITY 3.10

Table 3.1 lists the densities of various substances. Note that most solids and liquids have a density that is between 0.5 g/cm^3 and about 20 g/cm^3. But the densities of gases are only about 1/1,000 of the densities of solids and liquids.

Is the density of a substance always the same? Most substances expand when heated but their mass remains the same. Therefore, the density depends on the temperature, becoming less as the material expands and increases in volume. But, as we shall see in Chap. 9, the expansion is very small for solids and liquids and has little effect on the density. The situation is quite different with gases, which show a large thermal expansion. Moreover, we find it difficult to compress solids and liquids, but we can easily compress gases, as you know from pumping up a bicycle tire. Therefore, when measuring the density of a gas, we have to state the temperature and the pressure at which it was measured.

Table 3.1 Densities of Some Solids, Liquids, and Gases (in grams per cubic centimeter)

Osmium	22.5	Oak	0.6-0.9	
Platinum	21.4	Lithium	0.53	
Gold	19.3	Liquid helium (at −269°C)	0.15	
Mercury	13.6			
Lead	11.3	Liquid hydrogen (at −252°C)	0.07	
Copper	8.9			
Iron	7.8	Carbon dioxide	1.8×10^{-3}*	At atmospheric pressure and room temperature
Iodine	4.9	Oxygen	1.3×10^{-3}	
Aluminum	2.7	Air	1.2×10^{-3}	
Carbon tetrachloride	1.60	Nitrogen	1.2×10^{-3}	
Water	1.00	Helium	1.7×10^{-4}	
Ice	0.92	Hydrogen	8.4×10^{-5}	
Methyl alcohol	0.79	Air at 20 km (kilometers) altitude	9×10^{-5}	

*Small numbers less than 1 can, like large numbers, be most conveniently expressed in terms of powers of 10. For example, we write 0.1 as 10^{-1}, 0.01 as 10^{-2}, 0.001 as 10^{-3}, etc., using negative numbers for exponents.

If we have decimals like 0.002 we can write this first as 2×0.001 and then, in powers-of-10 notation, as 2×10^{-3}. Another example: $0.00009 = 9 \times 0.00001 = 9 \times 10^{-5}$. The negative exponent of the 10 tells how many places the decimal point must be moved to the left to give the correct value in regular notation.

18 Write the following numbers in powers-of-10 notation.
a) 0.001 0.1 0.0000001
b) 1/100 1/10,000

19 Write each of the following numbers as a number between 1 and 10 times the appropriate power of 10.
a) 0.006 0.000032 0.00000104
b) 6,000,000 63,700

20 Change the following numbers from powers-of-10 notation to ordinary notation.
a) 10^{-2} 10^{-5} 3.7×10^{-4}
b) 1.05×10^{-5} 3.71×10^{3}

21† A small beaker contains 50 cm^3 of liquid.
a) If the liquid were methyl alcohol, what would be its mass?
b) If the liquid were water, what would be its mass?

22† The densities in grams per cubic centimeter of various substances are listed below. Indicate which of the substances might be gas, liquid, or solid. (Refer to Table 3.1.)
(a) 0.0015 (b) 10.0 (c) 0.7 (d) 1.1 (e) 10^{-4}

23 Estimate the mass of air in an otherwise empty room that is the size of your classroom.

3.11 IDENTIFYING SUBSTANCES

We have looked for properties that can help us to distinguish between substances that appear to be the same. So far we have found three properties that do not depend on how much of a substance we have or on its shape. These properties are melting point, boiling point, and density.

Suppose we measured the melting points of two samples of matter and found them the same. If we then measured their boiling points and these were also the same, we might suspect that we had two samples of the same substance. We would not expect them to differ in their density or in any other properties. But, as Table 3.2 shows, we cannot depend on two properties alone to distinguish between substances. This is particularly true if the measurements are not highly accurate.

In Group 1 of the table we have substances with the same boiling points and nearly the same melting points. It would be hard to measure these two properties carefully enough to see that they are different substances, but a measurement of their densities would prove without question that they are different.

Table 3.2 Some Substances with Similar Properties

	Density (g/cm^3)	Melting point (°C)	Boiling point (°C)
Group 1			
Methyl acetate	0.93	−98	57
Acetone	0.79	−95	57
Group 2			
Isopropanol	0.79	−89	82
t-Butanol	0.79	26	83
Group 3			
Cycloheptane	0.81	−12	118
n-Butanol	0.81	−90	118
s-Butanol	0.81	−89	100

The names of the substances in this table are not important to us now, and you do not need to remember them. They are good examples of substances that we cannot tell apart unless we measure all three properties: density, melting point, and boiling point.

The substances in Group 2 have the same density and nearly the same boiling point but can be told apart by their different melting points.

If you compared only their densities, you might conclude that the three substances in Group 3 are the same. If you also measured their melting points, you would probably decide that the second and third substances in this group are the same. If you compared their densities and boiling points but not their melting points, which would you conclude are the same? In fact, all three substances in Group 3 are different. That is why they were given different names when first discovered.

There are not very many examples of substances that are nearly the same in two of these three properties and are different in the third. We would have to search long and hard to find two samples of matter that have the same density, melting point, and boiling point but that differ in some other property and are, in fact, samples of different substances. This means that if we can determine the densities, melting points, and boiling points of materials, we can distinguish between almost all substances.

In many cases, the melting point and the boiling point of a sample of matter can be measured easily in the laboratory. However, some substances have boiling points so high that it is difficult to get them hot enough to boil. Others have boiling points so low that it is difficult to get them cold enough to become liquid. For example, table salt boils at 1413°C. The same experimental difficulties come up when we try to determine the melting points of some substances. Grain alcohol melts at −117°C.

Suppose we have a sample of a newly made substance. We wish to find out whether it is truly a new substance, different from all others, or a substance already known but made in a new way. If its boiling and melting points are too high or too low to measure easily, we must look for other characteristic properties that might help to distinguish between similar substances.

24 Which of the substances listed in Table 3.2 are solids, which are liquids, and which are gases at (a) room temperature (20°C), (b) 50°C, (c) 100°C?

For Home, Desk, and Lab

25 Figure B shows a diagram of a double boiler. Why is such a double boiler used to cook food that is easily scorched?

FIG. B For prob. 25.

26 Object A has a mass of 500 g and a density of 5.0 g/cm^3; object B has a mass of 650 g and a density of 6.5 g/cm^3.
a) Which object would displace the most liquid?
b) Could object A and object B be made of the same substance?

27 A student measures the volume of a small aluminum ball by water displacement and then finds its mass on a balance. He finds that the sphere displaces 4.5 cm^3 of water and has a mass of 6.5 g.
a) What value does the student obtain for the density of aluminum?
b) How might you account for the difference between this value for the density of aluminum and the one given in Table 3.1?

28 A student has several specimens of substance A of different sizes and also of substance B of different sizes. She measures the masses and volumes of these specimens and plots the graphs shown in Fig. C. Which substance has the greater density? How do you know?

FIG. C For prob. 28.

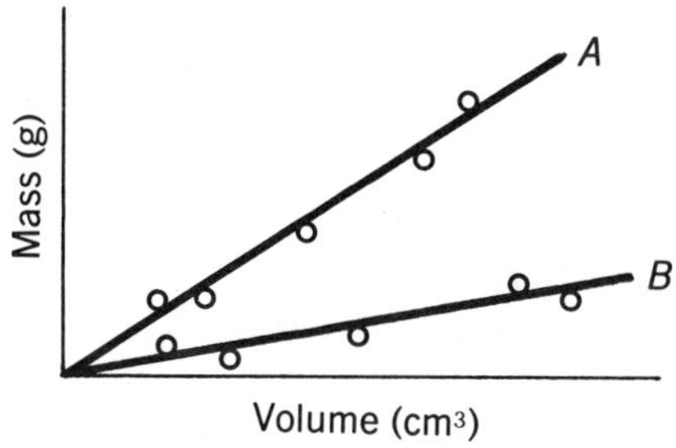

29 How would you determine the density of ice? Could you get the volume by melting the ice and measuring the volume of the resulting water?

30 How would you distinguish between unlabeled pint cartons of milk and of cream without breaking the seals?

31 Weight the end of a test tube with just enough sand so that it floats upright in water. With a pencil that marks on glass, mark the depth to which it sinks. How deep does it sink in alcohol? How could you use this device (called a "hydrometer") to measure the densities of unknown liquids?

32 In Table 3.1, why are the pressure and temperature stated for the densities of gases and not for the densities of solids and liquids?

33 The students in an IPS class in one of the coastal cities of the United States measured the boiling point of water and found that it was 100°C. On the same day the students in an IPS class in one of the mountain cities in the United States also measured the boiling point of water and found that it was 95°C. What can be inferred about the boiling point of water from these reports?

34 A cylinder is closed with a tight-fitting piston 30 cm from the end wall (Fig. D); it contains a gas that has a density of 1.2×10^{-3} g/cm^3. The piston is pushed in until it is 10 cm from the end wall; no gas escapes. What is the density of the compressed gas? What is your reasoning?

FIG. D For prob. 34.

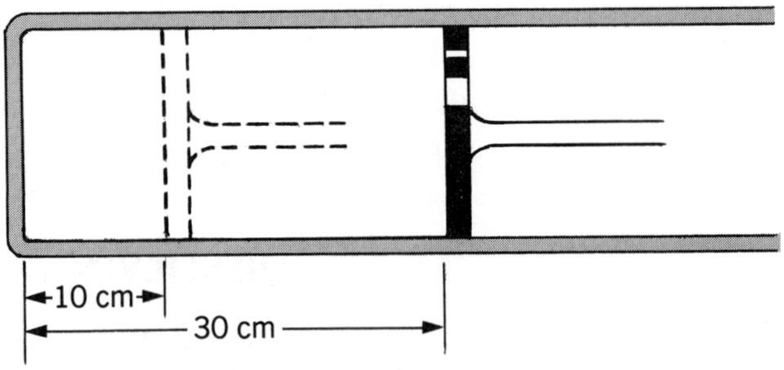

35 Does the density of air change when it is heated:

a) In an open bottle?

b) In a tightly stoppered bottle?

Solubility

4

EXPERIMENT DISSOLVING A SOLID IN WATER 4.1

You know from daily experience that sand and chalk do not dissolve in water but that sugar and table salt do. Of course, these are only qualitative observations. Are you sure that not even a tiny amount of chalk dissolves in a gallon of water? Can you dissolve as much salt in a glass of water as you wish?

Most solutions are colorless, and there is nothing to be seen once the solid is dissolved. To make things visible we shall begin our quantitative study of solutions with a solution that has color.

Place 5.0 cm^3 of water in one test tube and 20.0 cm^3 of water in another. Add 0.30 g of orange solid to each tube, and then stopper and shake thoroughly. Did all of the solid dissolve in both test tubes?

Is the color uniform in each tube? Do you think that each cubic centimeter of solution contains the same mass of dissolved material? Both tubes contain the same mass of dissolved material, but is the shade of the color the same in both tubes?

Will another 0.30 g of the orange solid dissolve in each of the test tubes? Try it but be patient! If the solution seems to cool as the solid dissolves, try to keep the temperature fairly constant by warming the test tube with your hand. Did 0.60 g of solid dissolve in 20.0 cm^3 as well as in 5.0 cm^3 of water? Is the color still uniform in each solution? Is the shade of the color the same in both solutions?

Add another 0.30 g of the orange solid to each solution. How much do you have in each test tube? What do you observe? How much orange solid do you think there would have to be in the 20.0 cm^3 test tube so that not all of it will dissolve? Test your prediction.

A solution in which no more solid can be dissolved is called a *saturated* solution.

4.2 CONCENTRATION

The uniformity of color in any one of the various solutions in the last experiment suggests that the solid dissolved uniformly. That is, a cubic centimeter of water in a given solution contained the same mass of orange material as any other cubic centimeter of water used to make the solution. For example, at the beginning of the experiment one test tube had $0.30\ g/5.0\ cm^3 = 0.060\ g/cm^3$ of material and the other test tube had $0.30\ g/20.0\ cm^3 = 0.015\ g/cm^3$.

The mass of solid dissolved per unit volume of liquid is called the *concentration* of the solution. This unit is the same as that of density, g/cm^3. However, in the case of density the mass and the volume refer to the same substance. In the case of concentration the mass refers to the dissolved solid (called the *solute*) and the volume refers to the liquid (called the *solvent*).

To avoid confusion and for other reasons of convenience, concentrations are often given in $g/100\ cm^3$.

A concentration of $0.015\ g/cm^3$ means that 0.015 g of solute is dissolved in 1 cm^3 of water. A volume of 100 cm^3 of water will, therefore, contain $100 \times 0.015\ g = 1.5\ g$ of solute. Therefore, the concentration of the solution in $g/100\ cm^3$ is $1.5\ g/100\ cm^3$. In general, to find the concentration of a solution in $g/100\ cm^3$, we multiply the concentration in g/cm^3 by 100.

---

1 For Expt. 4.1, calculate the concentration of the solutions in g/cm^3 and in $g/100\ cm^3$ after the addition of each sample of solid. GD

2 What can you say about the largest concentration you were able to make in the experiment?

3 A mass of 25.0 g of sugar is dissolved in 150 cm^3 of water. What is the concentration in $g/100\ cm^3$?

EXPERIMENT COMPARING THE CONCENTRATIONS OF SATURATED SOLUTIONS 4.3

From your results of dissolving the orange solid (Expt. 4.1) you know that you will reach a point where no more solute will dissolve in the solvent. The solution then has the largest possible concentration and is called a saturated solution.

To find the concentration of a saturated solution experimentally you could add a tiny amount of solid at a time and see whether it dissolves. A better method is to begin with a large mass of solid and shake until you judge that no more dissolves. Then you can pour off some of the clear liquid and determine the concentration.

Try dissolving 5 g of two solids in separate test tubes, each containing 5 cm^3 of water. Stopper the test tubes and shake them vigorously for several minutes until you have a saturated solution. If the tube cools during the process, keep it warm with your hand. Does one sample of solid appear to be more soluble in water than the other?

To find the concentrations of the two saturated solutions you can evaporate the liquid from a known mass of each solution. Subtracting the mass of the remaining dry solid from the mass of the solution will give you the mass and, therefore, the volume of the water of your sample. This will give you the data you need to calculate the concentration of the saturated solution. You can do the experiment for one solution while some of your classmates work with the other solution.

Pour almost all the saturated solution into a previously massed evaporating dish, being careful not to pour out so much solution that undissolved solid is carried over from the test tube into the dish.

After finding the total mass of dish and solution, you can slowly evaporate the saturated solution to dryness over a flame, as shown in Fig. 4.1, page 52, and find the mass of remaining solid. Be careful to heat the solution very slowly so that solid does not spatter out of the dish. Keep watching the dish, and move the flame away whenever the spattering begins.

What was the mass of the solid and the mass of the water in which it dissolved? What was the volume of the water? What was the concentration of each of the saturated solutions? Compare your results with those of your classmates.

FIG. 4.1
Evaporating a solution in an evaporating dish heated over an alcohol burner. If the liquid spatters, it can be heated more slowly by moving the burner to one side so that the flame heats only one edge of the dish.

The concentration of a saturated solution is called the *solubility*. In this experiment you found the solubilities of two substances in the same solvent, namely water.

Solubility is independent of the mass of the sample from which it is found. It is a characteristic property of the combination of the solute and the solvent.

---

4 Mario wishes to construct a table that lists the solubility in water of various substances. From various sources he finds the following data for solubilities at 0°C.

a)	**Boric acid**	**0.20 g in 10 cm^3 of water**
b)	**Bromine**	**25 g in 600 cm^3 of water**
c)	**Washing soda**	**220 g in 1,000 cm^3 of water**
d)	**Baking soda**	**24 g in 350 cm^3 of water**

What is the solubility of each substance in grams per 100 cm^3 of water?

5 From your answers to Question 4, find the largest mass of each substance that will dissolve in 60 cm^3 of water.

6 Suppose that 200 cm^3 of a saturated solution of potassium nitrate were left standing in an open beaker on your laboratory desk for three weeks. During this time most of the water evaporated. Would the mass of potassium nitrate dissolved in the solution change? Would the concentration of the potassium nitrate solution change during the three weeks?

EXPERIMENT THE EFFECT OF TEMPERATURE ON SOLUBILITY 4.4

In the last experiment you tried to keep the temperature of the solution constant (by warming the test tube with your hand if it cooled). How will the solubility of different substances—that is, the maximum mass of a solid that will dissolve in a given volume of liquid—be affected by the temperature of the liquid?

To find out, add 10 g of two solids to test tubes, each containing 10 cm^3 of water. Place both test tubes in a large beaker of water, and stir the solutions for several minutes until they are saturated. Now heat the beaker, stirring both solutions constantly, until the water in the beaker is nearly boiling. What do you observe? Do the solubilities of the substances appear to change equally or differently as the temperature of the water is increased? What do you predict will happen if you remove the burner and cool both test tubes together in a beaker of cold water? Try it.

——— ——— ———

Figure 4.2 shows the result of an experiment with potassium sulfate. The solubilities at different temperatures were measured by the same method you used in Expt. 4.3. The solubility is expressed as the mass in grams of the substance that is dissolved in 100 cm^3 of water to give a saturated solution or, to put it another way, the maximum mass of the substance that can be dissolved in 100 cm^3 of water.

Suppose we dissolve 20 g of potassium sulfate in 100 cm^3 of water at 80°C. If we now cool the solution to room temperature, 25°C, Fig. 4.2 shows that the water can hold only 12 g of potassium sulfate in solution at this temperature. Therefore, during the cooling process, small crystals of solid potassium sulfate begin to appear in the solution at about 70°C. As cooling continues, more crystals are produced in the solution and they sink to the bottom. A solid that crystallizes out of a saturated solution in this manner is called a "precipitate." As you can see from Fig. 4.2, the mass of potassium sulfate that will precipitate out of solution and collect at the bottom in this case will be 20 g − 12 g = 8 g.

Figure 4.3 shows the solubility as a function of temperature for several other common substances, all plotted together in the same graph. These curves clearly show that how the solubility of a substance changes with temperature is a characteristic property that

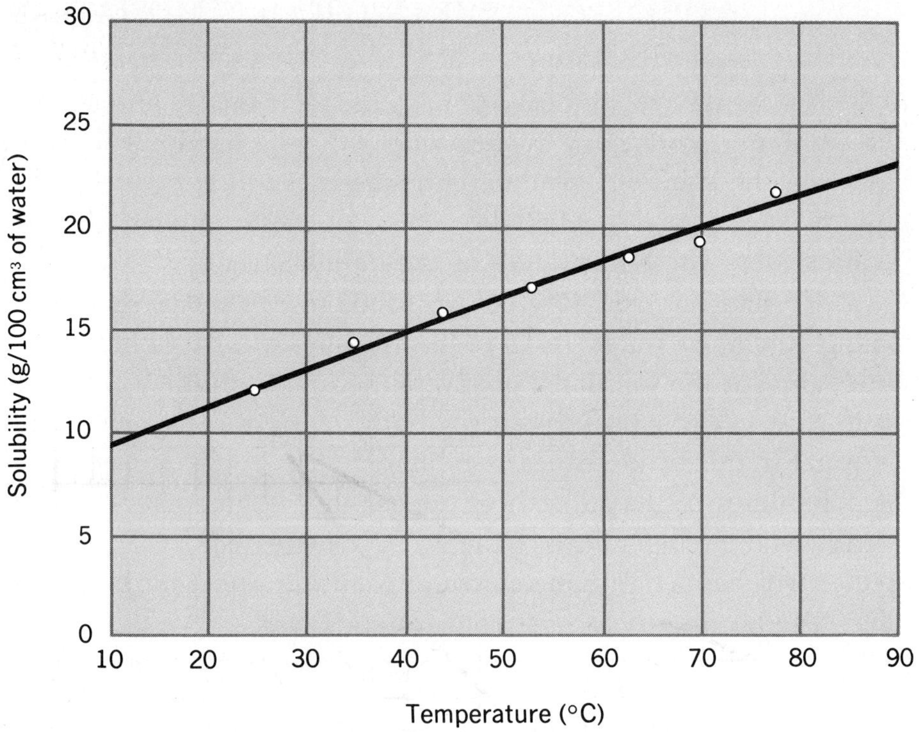

FIG. 4.2
A graph of the solubility of potassium sulfate as a function of temperature. The graph shows the maximum concentration of potassium sulfate solution at different temperatures.

can help to distinguish between different substances. You can see from the graph, for example, that the solubilities of potassium nitrate and sodium chloride (ordinary table salt) are very nearly the same at room temperature (about 20°C) but are widely different at high and low temperatures.

---

7 If you plotted the data from Fig. 4.2 on Fig. 4.3, where would the graph be found?

8 Suppose you have a saturated solution of potassium nitrate at room temperature. From Fig. 4.3 what do you predict will happen if you
a) heat the solution?
b) cool the solution?

9 What reason can you give for heating the solutions in Expt. 4.4 by immersing the test tubes in a beaker of hot water rather than heating the test tube directly over a burner flame?

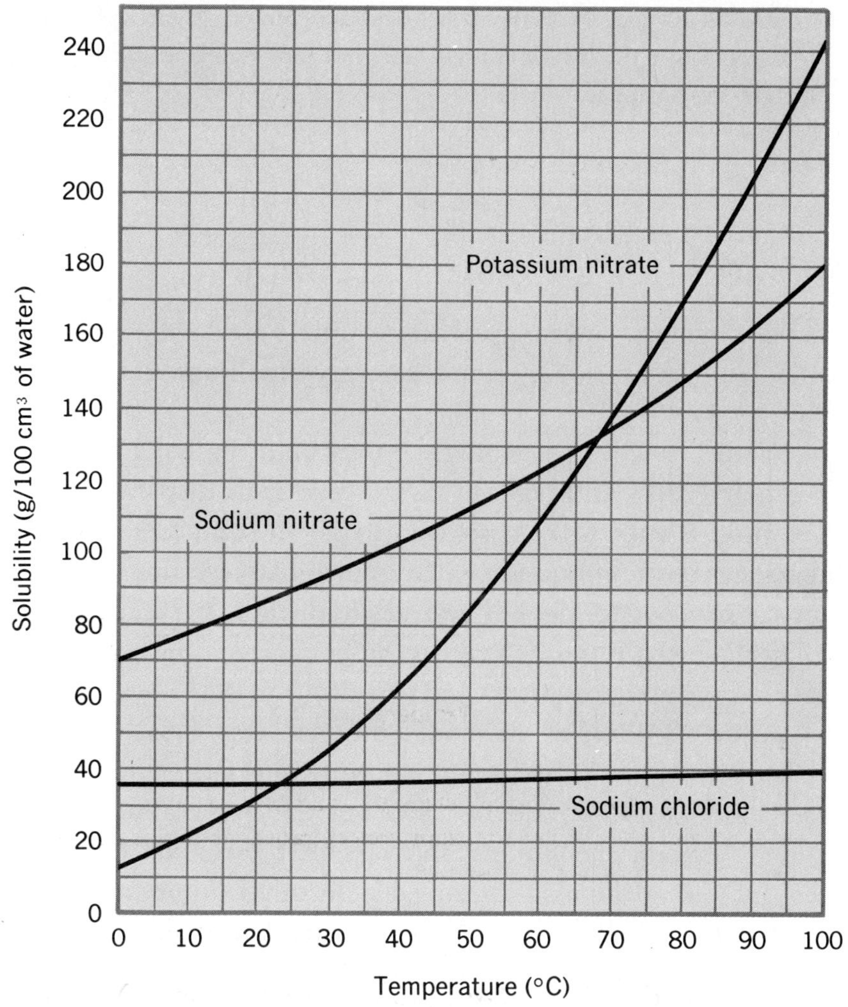

FIG. 4.3
Solubility curves of different substances dissolved in water. The maximum mass of material that will dissolve in 100 cm^3 of water is plotted along the vertical axis, and the temperature of the water along the horizontal axis.

10† What temperature is required to dissolve 110 g of sodium nitrate in 100 cm^3 of water?

11 a) If 20 g of sodium chloride is dissolved in 100 cm^3 of water at 20°C, is the solution saturated?
b) How do you know if a solution is saturated?

12† A mass of 30 g of potassium nitrate is dissolved in 100 cm^3 of water at 20°C. The solution is heated to 100°C. How many more grams of potassium nitrate must be added to saturate the solution?

13 A mass of 10 g of sodium nitrate is dissolved in 10 cm^3 of water at 80°C. As the solution is cooled, at what temperature should a precipitate first appear?

4.5 WOOD ALCOHOL AND GRAIN ALCOHOL

Most rocks, metals, and many other materials are so slightly soluble in water that we cannot measure the very small amounts that will dissolve.

Water, however, is not the only liquid. Perhaps some substances that hardly dissolve at all in water will dissolve easily in other liquids. If such is the case, we can use the different solubilities of substances in these liquids to distinguish between them. We shall first investigate wood alcohol and grain alcohol, two common liquids. Then we shall see if there are other solvents that will further increase our stock of tools for investigating matter.

Wood alcohol, as its name implies, was first made from wood. In fact, some of the liquid that you collected in your distillation of wood was wood alcohol. The ancient Syrians heated wood in order to obtain the liquids and tars that resulted. The watery liquids (including the alcohol mixed with other liquids) were used as solvents and as fuel for lamps. The tars were used to fill the seams in boats, to preserve wood against rot, and as mortar for bricks.

The method used by the Syrians in making these substances was the same as the one you used when you distilled wood. The apparatus they used was cruder. Lengths of wood were stacked closely in a dishlike depression in the top of a mound of earth. A drain ran from the middle of the depression to a collection pit. After the wood was covered with green branches and wet leaves, a fire was started inside the pile. As this fire smoldered, watery liquids and tars drained off from the pile and collected in the pit. Later it was discovered that the watery liquids could be separated, just as you separated them after you distilled wood. One of these liquids was wood alcohol.

Grain alcohol can be made by fermenting grains, such as corn, barley, and rye, and also by fermenting grapes and other fruits. Fermentation is the process that goes on naturally when fruit juices or damp grain are stored with little exposure to air. Gas will bubble out of the liquid. What remains boils at a temperature lower than the boiling point of water. As was discovered long before history was recorded, this liquid contained a new substance different from water. It was used as a beverage (with effects quite different from those of water) and as a medicine.

Sometime before the twelfth century, the wine from fermented grapes was first distilled, and the condensed liquid was described as the "water that burns." It was named *alcohol vini* or "essence of wine," and later came to be called "grain alcohol."

If we measure the densities of the alcohols we get from different grains and fruits, we find no difference between them. These alcohols also have the same boiling point and melting (or freezing) point. In fact, they are all the same substance. Similarly, the alcohols (or essences) produced from different kinds of wood are all the same: all have the same density, boiling point, and melting point. Since the distilled liquids obtained from wood and fermented grain are both called alcohol, you might think that they are the same substance. But from an examination of Table 4.1 we see that they are indeed different. Though their densities are nearly the same, their melting points and boiling points differ enough so that there can be no possibility that they are the same substance. Today, wood alcohol is called "methanol," and grain alcohol is called "ethanol."

Table 4.1 Some Characteristic Properties of the Most Common Alcohols

	Density (g/cm^3)	Melting point (°C)	Boiling point (°C)
Wood alcohol (methanol)	0.79	−98	64.7
Grain alcohol (ethanol)	0.79	−117	78.5

Both alcohols can be used as fuels. The fuel in your laboratory burner should contain mostly ethanol. Ethanol is the important ingredient in alcoholic beverages, but methanol is very poisonous. Some methanol, or other poison, is usually added to burner fuel to make it undrinkable. Both will dissolve many substances that are insoluble in water, and these alcohols have been used as solvents for centuries.

14 You distilled wood in Expt. 1.1. Where would you expect to find the methanol at the end of the stage of the experiment shown in Fig. 1.2? In Fig. 1.3?

4.6 EXPERIMENT METHANOL AS A SOLVENT

Sugar and citric acid look the same. They are both white. Sugar has a density of 1.59 g/cm^3, and citric acid has a density of 1.54 g/cm^3. These densities differ slightly, by only about 3 percent. You would find it difficult indeed to determine the volume of an irregular piece of either material to within 3 percent. The values that you would get for the densities of these substances could be no more accurate than your values for the volumes; and if either of these was in error by 3 percent or more, you could not be sure from your density determination whether the two pieces of material were the same substance or not. Furthermore, it is not easy to distinguish between these substances by their solubilities in water because both are about equally soluble in water.

See if you can distinguish between these substances by their solubility in methanol.

Do moth flakes dissolve in water? In methanol? What about magnesium and magnesium carbonate? Do they dissolve in water? In methanol?

--- --- ---

In this and later chapters we shall be using many substances with which you are not familiar. Why they have the names they do and what all their properties are may be interesting, but in most cases the answers to these questions are not important to us in this course. We shall use these materials only because they serve to illustrate clearly some particular properties that will help us in our investigation of matter.

15† Two solids appear to be the same and are both insoluble in methanol. A student, whose results are reliable to 5 percent, reports the solubilities of the two solids in water, as in the following table. Are these solids the same substance? Explain your answer.

Solid	Solubility ($g/100\ cm^3$)	
	0°C	100°C
A	73	180
B	76	230

16 a) Which of the following substances, *x*, *y*, and *z*, do you think are the same?
b) How might you test them further to make sure?

Substance	Density (g/cm^3)	Melting point (°C)	Boiling point (°C)	Solubility in water at 20°C ($g/100\ cm^3$)	Solubility in methanol at 20°C
x	1.63	80	327	20	insoluble
y	1.63	81	326	19	insoluble
z	1.62	60	310	156	insoluble

SULFURIC ACID 4.7

Another solvent useful in distinguishing between different substances was first produced more than a thousand years ago by heating a soft rock called "green vitriol." The vapors produced by heating the rock were cooled and condensed to form an oily liquid, which was called oil of vitriol. Its modern name is "sulfuric acid." The ancient method of producing this acid has been abandoned in favor of a far more effective process of manufacture that starts with sulfur. As a result the acid is readily available in large quantities. It is of great importance in industry, being used in the manufacture of very many other substances.

One of the useful properties of sulfuric acid is its ability to dissolve many substances that will not dissolve in water. In many cases in which a substance is being dissolved in sulfuric acid a gas is given off. We shall investigate two such gases in the next experiment.

EXPERIMENT TWO GASES 4.8

One of the substances that produces a gas when dissolved in sulfuric acid is magnesium. To collect the gas you can use the same apparatus as in Expt. 3.9. (**CAUTION:** Sulfuric acid is highly corrosive, and so try not to spill it or get any on your hands, clothing, or books. If you do, wash it off immediately with water.) Use about half a test tube of acid and five 7-cm lengths of magnesium ribbon to produce several test tubes of gas. You should discard the gas collected in the first test tube. Why?

Will the gas burn? Should you hold the test tube of gas upside down or right side up after you remove it from the water to test it? Try lighting tubes of the gas both ways, using first a burning splint and then a glowing one in each case. Is it more dense or less dense than air? Try bubbling a little of the gas into limewater (Fig. 4.4).

Repeat the experiment, using magnesium carbonate instead of magnesium.

4.9 HYDROGEN

The gas you made by dissolving magnesium in sulfuric acid was first produced many centuries ago by the action of sulfuric acid on

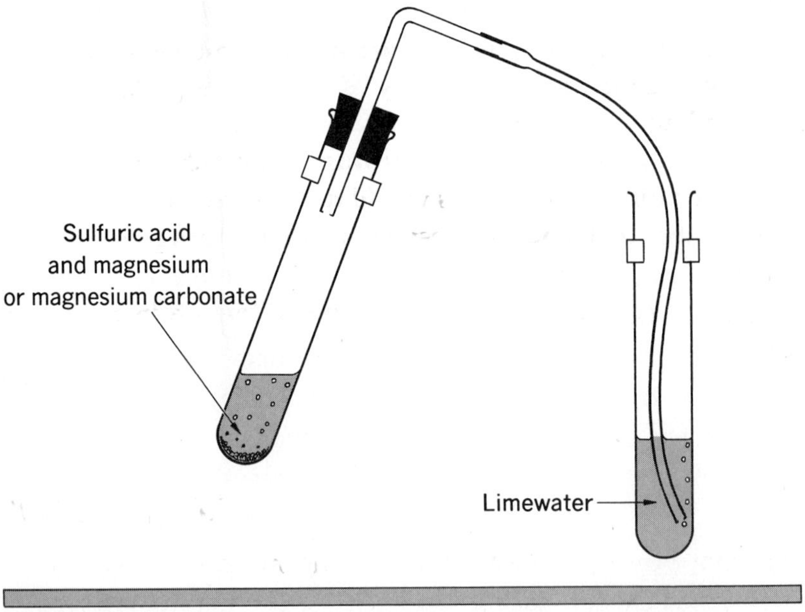

FIG. 4.4
The limewater test. Gas from the test tube on the left is bubbled through limewater in the one on the right.

metals. It was called "inflammable air" because it burned. Later, it was discovered that when inflammable air burns, it produces another gas, which can easily be condensed to a liquid. This liquid was found to boil at 100°C, freeze at 0°C, have a density of 1.00 g/cm^3, and dissolve the same substances that water does. In fact, all the properties of this liquid were the same as those of water, and so it was considered highly unlikely that it could be anything else. For this reason, about 200 years ago the gas that produced water when burned was given the name "hydrogen." "Hydro" is a prefix that means "water"; "gen" is a suffix meaning "producing" or "generating."

However, there are other gases that also burn and give water and, like hydrogen, are less dense than air. For many years these gases were thought to be the same as hydrogen. Marsh gas, now called "methane," is an example, and it has been known for thousands of years. It is often produced by decaying vegetable matter at the bottoms of lakes or marshes and slowly bubbles up to the surface of the water. But this gas does not have nearly so low a density as hydrogen, and so it must be a different substance. It was not until the density of gases could be measured with reasonable accuracy that marsh gas was found to be a different substance from hydrogen.

---

17 A student might try to find the density of hydrogen by the method you used for finding the density of a gas in Expt. 3.9, where you collected about 400 cm^3 of gas.

a) What would be the mass of this volume of hydrogen?

b) What difference in mass would there be between the total mass of test tube, acid, and magnesium at the start and the total mass of test tube and contents at the end of the action?

c) Could you accurately measure this difference in mass on your balance?

CARBON DIOXIDE 4.10

The gas that you produced by dissolving magnesium carbonate in sulfuric acid is called "carbon dioxide." It is a part of the air we exhale at every breath and is produced by a variety of substances when they are placed in sulfuric acid. The density of carbon dioxide is 1.8×10^{-3} g/cm^3, which is much greater than that of air. Since carbon dioxide does not burn, it makes a good fire extinguisher. It simply smothers the fire like a blanket.

Solid carbon dioxide, called "Dry Ice," evaporates directly into carbon dioxide gas without first turning into a liquid. But, like melting ice, it stays cold until all the solid is gone. Ice melts at 0°C, but Dry Ice evaporates at a much lower temperature (at −78.5°C). It remains at this temperature as it evaporates, and for this reason it is often used to cool things down to very low temperatures.

---

18 Describe how you would show that carbon dioxide is produced in a candle flame.

19 Carbon dioxide is contained in many common fire extinguishers. Give two reasons why it is a good fire extinguisher.

4.11 THE SOLUBILITY OF GASES

If the gases we have investigated so far were very soluble in water, we could not conveniently collect them over water. They would simply dissolve as fast as we produced them. In fact, these gases are so insoluble that it would be difficult to measure the amount that dissolves and then distinguish between them by their solubility. Are there any gases that are soluble enough that we can easily distinguish them by their solubility? Yes; an example is ammonia, a common enough substance, whose solution in water is used as a household cleaner.

20† In certain shallow parts of Long Island Sound fish have been found dead of oxygen starvation on extremely hot days, though at other times fish live there very happily. What property of oxygen do these facts suggest?

21 Experiment 3.9 was performed twice, but with one tablet and 15 cm^3 of water in each case. When the rubber tube was placed as shown in Fig. 3.6, 435 cm^3 of gas was collected. When the tube reached only slightly beyond the mouth of the bottle, only 370 cm^3 of gas was collected. The change in mass of the reactants was the same in both cases.

a) What volume of gas dissolved in the water?

b) Use Table 3.1 (page 43) to find what mass of gas dissolved in the water.

22 The following table shows the solubility of carbon dioxide at atmospheric pressure at various temperatures.

Solubility (g/100 cm^3)	Temperature (°C)
0.34	0
0.24	10
0.18	20
0.14	30
0.12	40
0.10	50
0.086	60

a) Draw a graph from these data.
b) Find, from your graph, how much carbon dioxide will dissolve in 100 cm^3 of water at 25°C.
c) Use Table 3.1, page 43, to find the volume of carbon dioxide that will dissolve in 100 cm^3 of water at room temperature (20°C).
d) How much carbon dioxide will dissolve in 100 cm^3 of water at 95°C?

EXPERIMENT THE SOLUBILITY OF AMMONIA GAS 4.12

Figure 4.5, page 64, shows how ammonia gas can be produced by slowly heating a water solution of ammonia. The gas can then be collected in a dry test tube. What can you conclude about the solubility of the gas as the temperature of the solution is raised? From the method of collecting the gas, what can you conclude about its density?

Collect a test tube of the gas. (Don't try to smell it directly; it is very irritating.) Then, very slowly, remove the tube and close it with a stopper. Place it, mouth downward, in a beaker of water, and then remove the stopper.

Even though most gases are nearly insoluble in water, they dissolve slightly, and it is fortunate that they do. Fish, for example, obtain the oxygen they need from the oxygen dissolved in the water around them. Carbon dioxide dissolved in sea, lake, and stream water provides one of the necessary materials used by shellfish to form their shells.

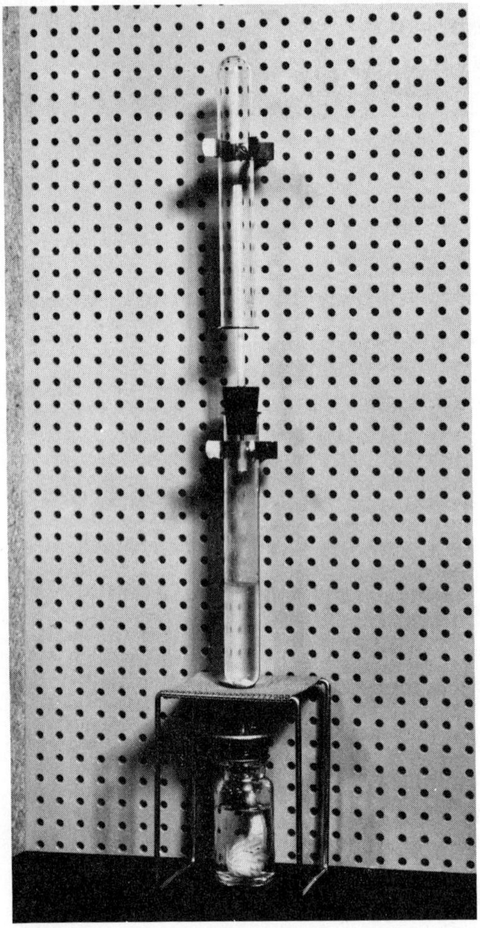

FIG. 4.5
Collecting ammonia gas produced by heating a water solution of the gas.

23† If the tube of ammonia gas had been inverted and placed in water saturated with ammonia, would the liquid have risen in the test tube as it did in this experiment?

4.13 OTHER SOLVENTS

Except for water and the two alcohols we have discussed in this chapter, sulfuric acid was the first and most widely used solvent. Later it was found that additional solvents could be prepared by heating green vitriol with common salt and by heating it with potassium nitrate. The solvents thus obtained are called hydrochloric acid and nitric acid respectively. Today these substances are made more effectively by other processes.

You can now see how much we have enlarged our collection of tools for distinguishing between substances. We do not always have to measure density, melting point, or boiling point. For example, suppose we have two test tubes of colorless liquids and we place a piece of copper in each. If the copper dissolves in one but not in the other, we know the two liquids are different. Similarly, if we have two samples of white crystals that dissolve equally well in one solvent but do not dissolve equally well in another solvent, then they are different substances. You saw an example of this in the case of sugar and citric acid.

For Home, Desk, and Lab

24 **The solubility of chalk in water is 10^{-3} g/100 cm^3. How much water would be necessary to dissolve a piece whose mass is 5 g?**

25 **Each of four test tubes contains 10 cm^3 of water at 25°C. The following masses of an unknown solid are placed in the test tubes: 4 g in the first, 8 g in the second, 12 g in the third, and 16 g in the fourth. After the tubes are shaken, it is observed that all of the solid has dissolved in the first two tubes, but that there is undissolved solid in the remaining two tubes.**

a) What is the concentration of the solid in each of the first two tubes?

b) What can you say about the concentration of the solid in the second two tubes?

26 **a) Which of the substances shown in Fig. 4.3 could be the unknown solid of Prob. 25?**

b) If the unknown is indeed the substance you named in answer to part (a), what will happen if the solution in each test tube is cooled to 10°C?

27 **The solubility of a substance in water was found to be 5 g/100 cm^3 at 25°C, 10 g/100 cm^3 at 50°C, and 15 g/100 cm^3 at 75°C. What would you expect its solubility to be at 60°C? At 100°C? Explain.**

28 **In many localities, after a kettle has been used for some time for boiling water, a flaky solid appears on the inside-bottom and on the sides of the kettle where the water has been in contact. How do you account for the presence of this so-called "boiler scale"?**

29 Your experiment with moth flakes (Expt. 4.6) showed that this substance was insoluble in water but dissolved readily in methanol.
a) Predict the effect of adding water to a methanol solution of moth flakes. Try it.
b) Sugar proved to be almost insoluble in methanol but dissolved readily in water. Predict the effect of adding methanol to a solution of sugar in water. Try this too.

30 In Expt. 4.8 you found that magnesium metal would dissolve in sulfuric acid.
a) Does this observation enable you to predict with certainty that all metals will dissolve in sulfuric acid?
b) Try dissolving other metals, such as copper, zinc, lead, and aluminum, in sulfuric acid.

31 When magnesium carbonate is placed in sulfuric acid, a gas is produced whose properties you studied in Expt. 4.8. When you place washing soda in hydrochloric acid, you also get a gas. What would you do with this gas to determine whether it is the same gas you got from dissolving magnesium carbonate in sulfuric acid?

32 Three samples of gas are tested for characteristic properties. Sample ***A*** does not turn limewater milky, is less dense than air, and burns. Sample ***B*** turns limewater milky, is more dense than air, and burns. Sample ***C*** turns limewater milky, is less dense than air, and burns. What can you conclude about these samples of gas?

33 A fizzing tablet is dissolved in 10 cm^3 of water, and the gas is collected as in Expt. 3.9. The volume of gas collected is 450 cm^3. When 50 cm^3 of water is used, the volume of gas collected is 405 cm^3. The tube was all the way up in the bottle in both cases.
a) Why do you think the volume is less?
b) Would this make a difference in the density calculation?

34 If you have a certain amount of a solid to dissolve in water, you usually can hasten the process by (a) stirring the water, (b) crushing the solid into smaller particles, and (c) heating the water. Why do you think each of these steps is effective in making the solid dissolve faster?

The Separation of Substances

5

In this chapter, we shall use the characteristic properties we have studied to work out a variety of methods for separating mixtures of different substances, whether these mixtures be gases, liquids, or solids. By learning how to separate many kinds of mixtures, we hope that we shall also come closer to understanding how simple substances are put together to form the many complex materials that are all around us. We shall start with a careful investigation of what happens when we distill a mixture of liquids.

EXPERIMENT FRACTIONAL DISTILLATION 5.1

In this experiment you will determine some of the properties of a mixture of liquids. Then you will distill the mixture and examine the properties of the fractions to see if you succeeded in separating the liquids that made up the original mixture.

Part A

Can you tell just by looking that the liquid is a mixture? Does it have an odor? Dip a small piece of paper in it, and try lighting the liquid on the paper with a match. Does the liquid burn? Find its density. Does sugar dissolve in the liquid?

Part B

Use the apparatus shown in Fig. 5.1 to distill 5 cm^3 of the mixture. Use a single collecting tube, and heat the liquid just enough to keep

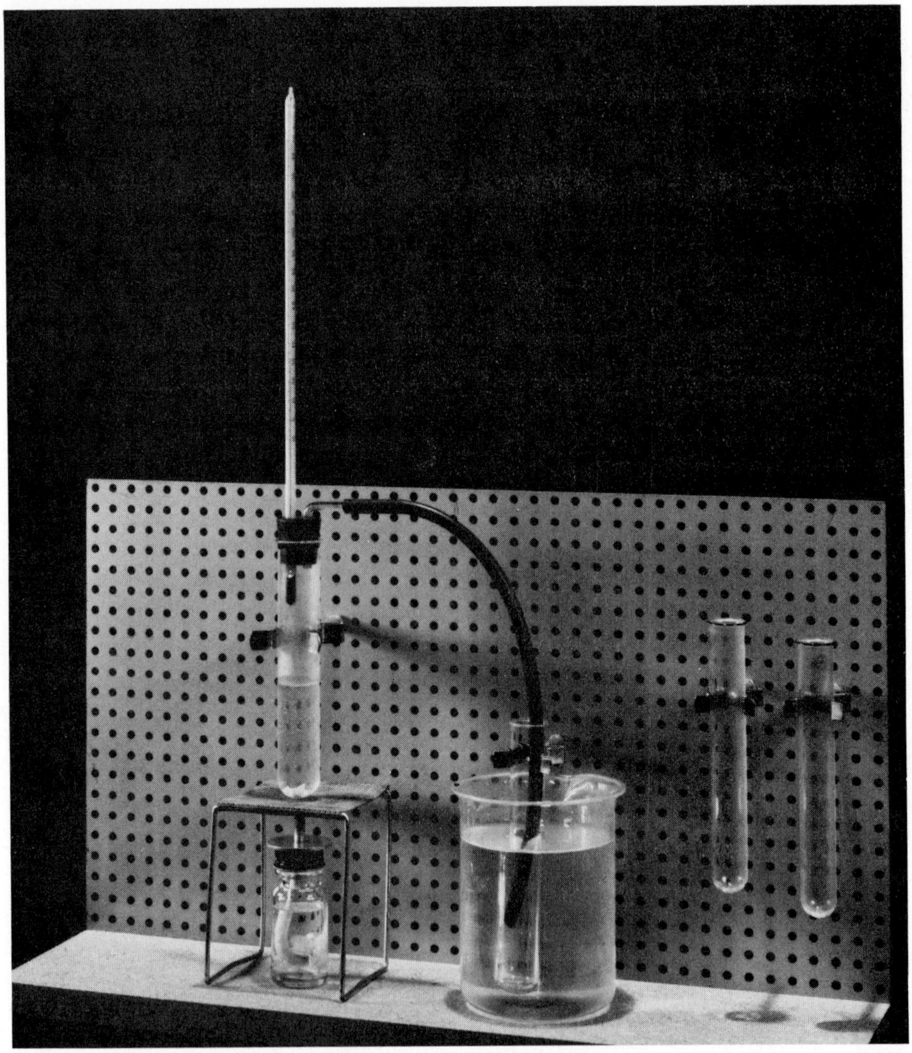

FIG. 5.1

Apparatus for the fractional distillation of a liquid. The thermometer bulb is close to the top of the test tube so that it measures the temperature of the vapor that condenses in the outlet tube. If there is more than one liquid in the boiling mixture, most of the high-boiling-point liquids will condense and flow back down the test-tube walls before they reach the upper part of the test tube. The thermometer in this apparatus is not used to measure the boiling temperature of the liquid but serves to show when collecting tubes should be changed to receive different fractions.

it boiling. Record the temperature of the *vapor* from the boiling liquid every half minute while it distills. Continue to boil the liquid almost to dryness.

Make a graph of the boiling temperature as a function of the time. What do you conclude from your graph about the number of fractions you should collect to separate the different substances in the mixture? At what temperatures should you shift from one collecting tube to another? Indicate on your graph the temperatures at which you decided to change collecting tubes.

Part C

Now fractionally distill about 25 cm^3 of the liquid. Label the test tubes containing the fractions so that you can keep track of them throughout the rest of the experiment.

Test each of the fractions for smell and flammability.

Part D

What is the density of fraction 1? Distill the fraction into a single test tube, recording the boiling temperature every half minute until the fraction has nearly boiled away. Draw a boiling-point graph for fraction 1.

Does sugar dissolve in fraction 1?

Part E

Repeat Part D for each of the other fractions.

Part F

Summarize your findings and compare the smell, flammability, density, ability to dissolve sugar, and boiling point of each of the fractions and of the original mixture. What do you conclude about the composition of the fractions? Can you identify the substances in the mixture (see Table 3.2, page 45)? What other tests might you make to help identify these substances? What do you think would happen if you were to redistill each of the fractions separately into three fractions of equal volume?

1† In what characteristic property must two liquids differ before we can consider separating a mixture of them by fractional distillation?

2 **The temperature-time graph shown in Fig. A was made during the fractional distillation of a mixture of two liquids *A* and *B*, and fractions were collected during the time intervals I, II, III, and IV. Liquid *A* has a higher boiling point than liquid *B*. What liquid or liquids were collected during each of the time intervals?**

FIG. A For prob. 2.

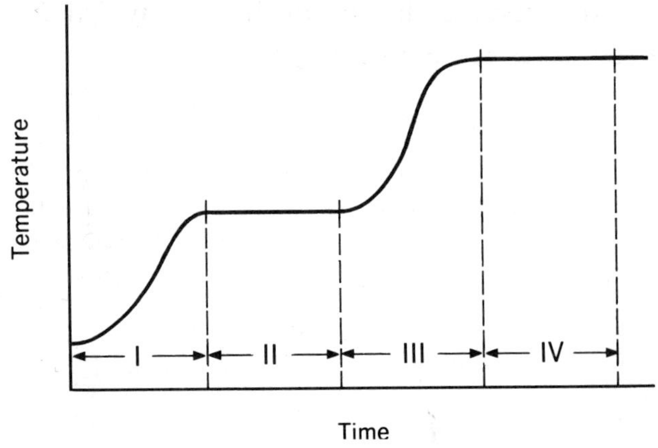

3 **Carlos boiled a liquid and recorded the temperature at 1-min intervals until the liquid had nearly boiled away. How do you explain the shape of the curve he got? (See Fig. B.)**

FIG. B For prob. 3.

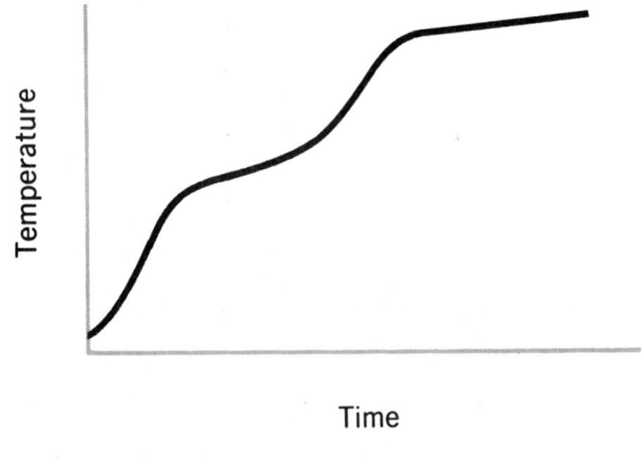

5.2 PETROLEUM

It is not always easy to separate a mixture of liquids into pure substances by fractional distillation. If the boiling points of the substances in a mixture are nearly the same, they will all boil off together. In a liquid mixture containing many substances, some are sure to have boiling points close together. When we distill such a mixture, each fraction we get is made up of a number of different substances whose boiling points are close together. The first part to condense in one fraction may contain some of the substances that condensed in the last part of the previous fraction. However, the fractions that are more widely separated in boiling range are less likely to contain the same substances. Petroleum is another example of such a mixture; the composition of typical fractions distilled from petroleum is shown in Table 5.1.

Table 5.1 A Few of the Substances Found in Petroleum

Substance	Density at 0°C (g/cm^3)	Freezing point (°C)	Boiling point (°C)	Common products of petroleum: Fuel gas	Gasoline	Kerosine
Methane	7.16×10^{-4}	−182.5	−161	X		
Ethane	1.35×10^{-3}	−183	−88	X		
Propane	2.02×10^{-3}	−190	−43	X		
Butane	2.68×10^{-3}	−138	−0.5	X		
Pentane	0.626	−129	36			
Hexane	0.660	−94	69		X	
Heptane	0.684	−90	98		X	
Octane	0.703	−57	125		X	
Nonane	0.722	−51	151		X	X
Decane	0.730	−30	174		X	X
Undecane	0.741	−26	196			X
Dodecane	0.750	−10	216			X
Tridecane	0.755	−5.5	236			X
Tetradecane	0.765	5.5	254			X
Pentadecane	0.776	10	271			X
Hexadecane	0.773	18	287			X

There are many more substances in the above products and also in the higher-boiling-point fractions not listed in the table—fractions such as fuel oils, lubricating oils, waxes, asphalt, and coke (mostly carbon).

Petroleum is believed to be produced naturally from dead animal and vegetable matter at the bottoms of shallow seas and swamps. When tiny plants and animals die in the sea they settle slowly to the bottom, where they become trapped in mud and sand. This sediment of mud, sand, and dead organisms slowly becomes thicker and thicker. In a million years, it may become thousands of feet deep. Such layers of sediment are very heavy, and the lower layers are compressed so much that they turn into rock layers. During this time, some of the body tissue of the entrapped organisms is changed into a viscous, sticky liquid that is a mixture of many thousands of different substances. This liquid is called "petroleum" or "crude oil." It is slowly squeezed out of the sediment in which it forms and eventually spreads through porous rock layers like water in a sponge.

In the course of more millions of years, the slow but ever-changing crust of the earth—buckling in some places, rising in others, and sinking in still others—moves and compresses the rock layers that were on the ocean bottom. If the porous oil-bearing rock is covered by a layer of hard, nonporous rock that has been bent into a dome or arch, as shown in Fig. 5.2, the oil will be trapped and cannot by itself squeeze to the surface. If it were not trapped, much of it would soon wash away and be lost. Most of the petroleum in the earth's crust is stored by nature under formations of nonporous

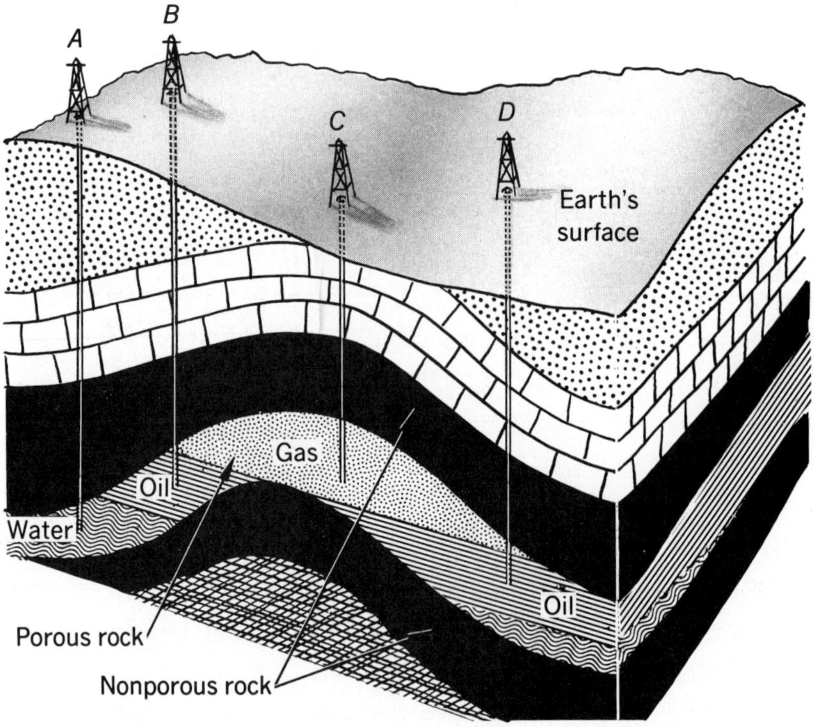

FIG. 5.2
A cross section of the earth's crust, showing how oil and natural gas are trapped in a porous rock layer by nonporous rock layers above and below. Note that well *A* produces only water, and well *C* produces only gas.

rock, which trap the liquids below them. As Fig. 5.2 shows, salt water (from the sea) and natural gas (the low-boiling-point substances in petroleum) are often trapped along with the oil. The nonporous "cap rock" may be thousands of feet thick. It is expensive and difficult to drill through all this rock to get to the petroleum below, and it is not easy to predict where oil is trapped. Deep and expensive wells often fail to reach oil or gas. Some wells produce nothing but salt water, whereas others remain always dry.

Petroleum was first discovered where it seeped to the surface in shallow pools. Once exposed to the air, some of the lower-boiling-point substances slowly evaporated, leaving tarry, almost solid, asphalt behind. These tars, as well as the liquid petroleum, were used for many of the same purposes in the ancient world as the tars and watery substances obtained from the distillation of wood.

One of the ancient methods used to distill crude oil consisted of heating the oil in a copper urn with a wool "sponge" at the narrow mouth of the vessel. The vapors condensed in the sponge, which was squeezed out into containers from time to time. A variation of this method made use of a heavy wick of wool that led from the mouth of the urn into a collecting vessel. Such a wick was a crude form of condenser.

The widespread use of kerosine lamps a little over 100 years ago—and, more recently, of gasoline engines—created a demand for petroleum. This led to improved methods of locating oil and drilling wells. Better apparatus was also developed for fractionally distilling petroleum on a large scale. This apparatus gives the useful fractions with which we are familiar—gasoline, kerosine, diesel fuel, heating and lubricating oils, paraffin, and asphalt—as well as less familiar fractions used in industry.

---

4 A sample of crude oil is boiled for several minutes. What change takes place in its density? (See Table 5.1.)

THE DIRECT SEPARATION OF SOLIDS FROM LIQUIDS 5.3

We have seen that fractional distillation can be used to separate a mixture of liquids. Sometimes we have a solid mixed with a liquid, and we wish to separate the solid from the liquid. If the solid does not dissolve in the liquid, the solid may settle to the bottom of a container. We can then pour off the liquid and dry the solid. (A mixture of sand and water is an example.) If the insoluble solid is less dense than the liquid, it will float on top and can be skimmed off and dried (for instance, a mixture of sawdust and water). Such simple methods will not work well if the solid does not separate from the liquid by itself in a short time. In such cases we must use other methods. One method is to filter the mixture.

If the solid is dissolved in the liquid, we can boil the solution until there is only a dry solid left. The liquid is condensed and collected separately.

5† How could drinking water be obtained from seawater?

6† Suppose you have a barrel filled with a mixture of sand, small pebbles, and stones. How would you separate the components of this mixture?

5.4 EXPERIMENT SEPARATION OF A MIXTURE OF SOLIDS

How can solids be separated from one another? If they are made up of large particles of different colors, we could, perhaps, separate them by using a magnifying glass and a pair of tweezers. This would be like separating blue marbles from white, or sugar cubes from ball bearings, and it would be a tedious process at best. For solids that are ground up very finely and well mixed, this method would not work at all.

Examine the mixture of solids supplied by your teacher. Do you think you could separate the substances by using the method of the magnifying glass and tweezers? If one solid is soluble in water and the other is not, you can separate them easily by taking advantage of the difference in solubilities.

Try to separate the two substances by dissolving one and separating it from the other by filtering. You can do this in the following way: Put about 1.5 g of the mixture in a test tube, and add 5 cm^3 of water. Stopper the test tube, and shake it for several minutes. Do you think either substance dissolved? To find out, filter out the undissolved material, as shown in Fig. 5.3. Wash the precipitate left on the filter paper by pouring an additional 10 cm^3 of water into

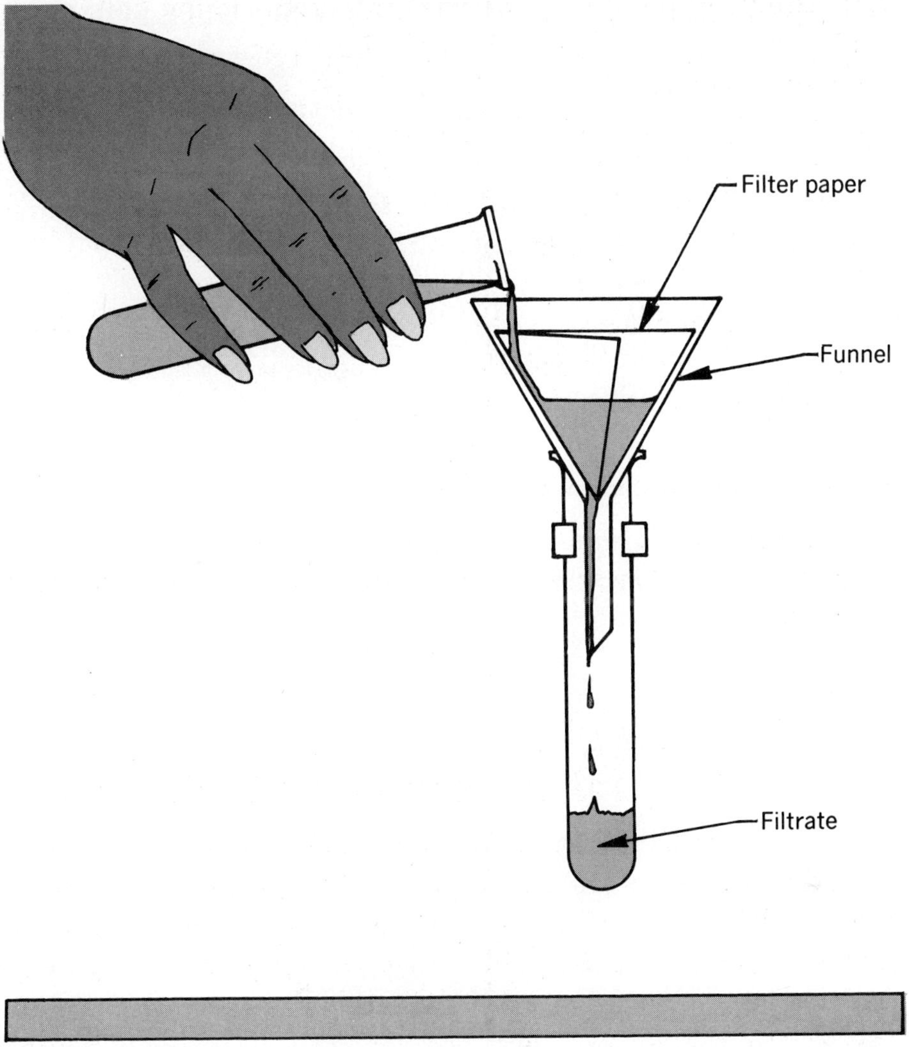

FIG. 5.3
Filtering a liquid. The filter paper is folded into a cone, which fits snugly into the funnel.

the funnel. You can now put about 5 cm^3 of the clear liquid, the filtrate, into an evaporating dish and boil it to dryness. Have the two substances been separated?

7† In what characteristic property must two solids differ if they are to be separated merely by dissolving at room temperature and filtering?

8 Much salt is obtained from salt mines, in which great masses of salt occur mixed with insoluble earthy impurities. What steps could be taken to purify the salt?

5.5 THE SEPARATION OF A MIXTURE OF SOLUBLE SOLIDS

In the previous experiment you were able to separate two solids because one of them was soluble in water and the other was not. Suppose that you have a mixture containing 5 g of sodium chloride and 5 g of potassium nitrate. Both of these substances readily dissolve in water. In fact, as you see from Fig. 5.4, at room temperature they nearly have the same solubility. Therefore, if we completely dissolve the 5 g of potassium nitrate in water, at room temperature, all of the sodium chloride will also dissolve. How can these substances be separated?

Figure 5.4 shows that the solubility of sodium chloride hardly changes with temperature whereas the solubility of potassium

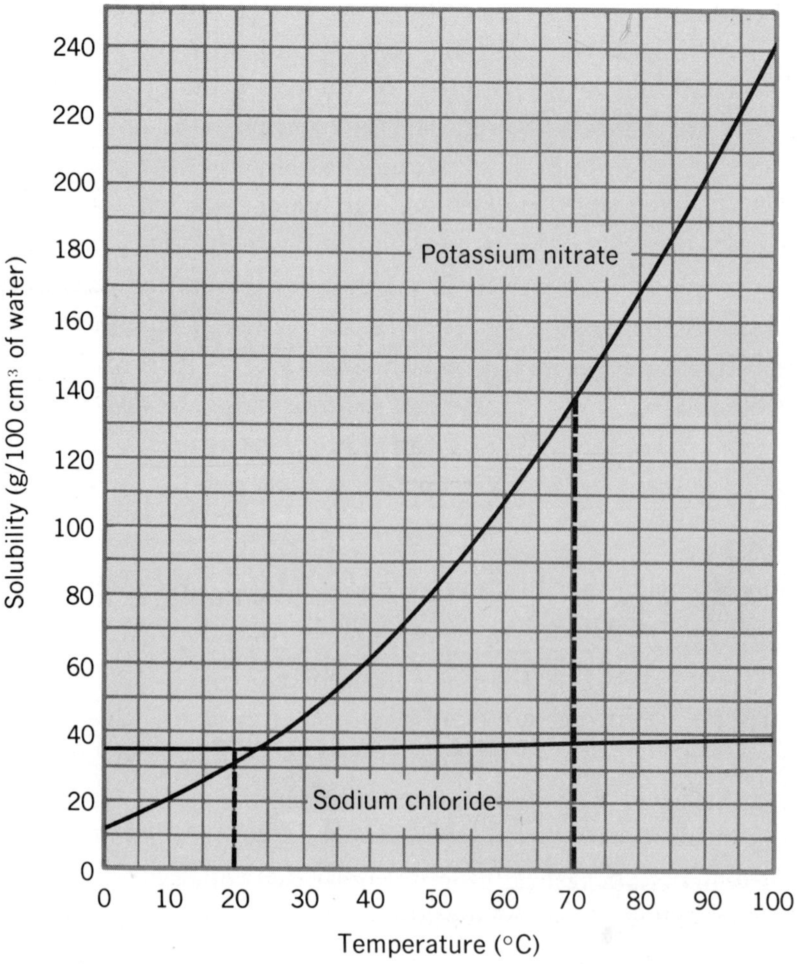

FIG. 5.4
Solubility curves of sodium chloride and potassium nitrate.

nitrate rises sharply with temperature. We can make use of the effect of temperature on the solubility of these two substances to separate them.

Note that at 70°C, for example, 138 g of potassium nitrate will dissolve in 100 cm^3 of water. Therefore, the amount of water needed to completely dissolve only 5.0 g of potassium nitrate at this temperature is 100 $cm^3 \times 5.0$ g/138 g $= 3.6$ cm^3.

How much sodium chloride will dissolve in 3.6 cm^3 of water at 70°C? To answer this question, first note from Fig. 5.4 that 38 g of sodium chloride will dissolve in 100 cm^3 of water at 70°C. Therefore, only 38 g $\times$ 3.6 cm^3/100 $cm^3 = 1.4$ g of sodium chloride will dissolve in 3.6 cm^3 of water.

Thus, if we add a mixture of 5.0 g of potassium nitrate and 5.0 g sodium chloride to 3.6 cm^3 of water, all the potassium nitrate and a small amount of sodium chloride will dissolve at 70°C. Most of the sodium chloride will remain as a solid at the bottom of the test tube. Therefore filtering the hot solution will leave most of the sodium chloride on the filter paper. The solution that passes through the filter paper will contain all the potassium nitrate and some sodium chloride.

What will happen if we let the filtered solution cool to room temperature? Referring again to Fig. 5.4, we see that practically all of the dissolved sodium chloride and some potassium nitrate will remain in the solution. However, most of the potassium nitrate will precipitate out as a solid and can then be separated from the solution by filtration. This entire process is summarized in Fig. 5.5.

---

9 a) How much sodium chloride does not dissolve in the 3.6 cm^3 of water at 70°C?

b) How much sodium chloride remains in solution after cooling to room temperature?

c) How much solid sodium chloride is recovered at the end of the experiment?

10 a) How much potassium nitrate remains in solution at room temperature at the end of the separation?

b) How much potassium nitrate can be recovered in solid form?

11 Suppose you wish to separate more of the two solids remaining in solution at the end of the experiment. How would you proceed to do this?

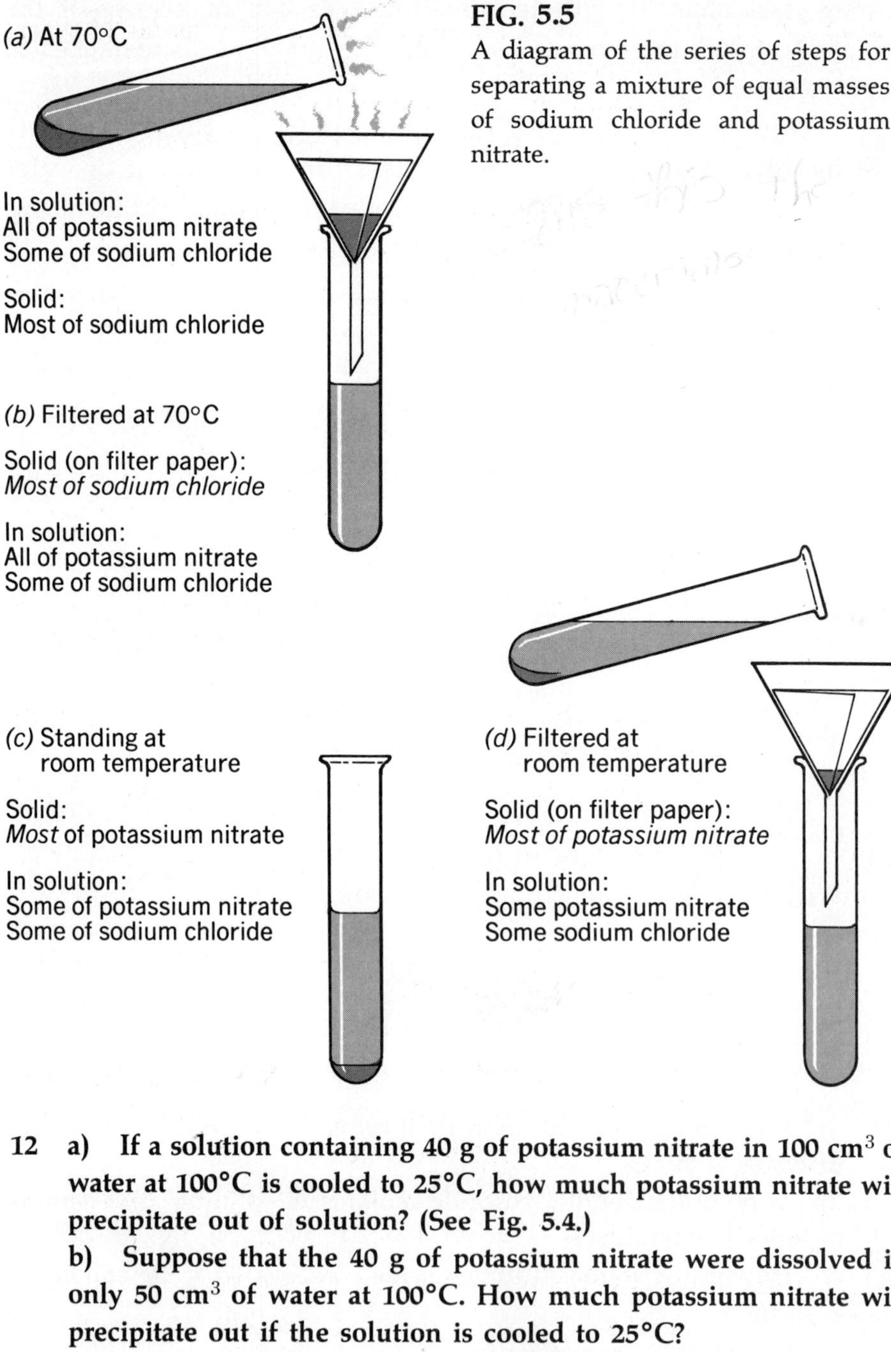

FIG. 5.5
A diagram of the series of steps for separating a mixture of equal masses of sodium chloride and potassium nitrate.

12 **a) If a solution containing 40 g of potassium nitrate in 100 cm^3 of water at 100°C is cooled to 25°C, how much potassium nitrate will precipitate out of solution? (See Fig. 5.4.)**
b) Suppose that the 40 g of potassium nitrate were dissolved in only 50 cm^3 of water at 100°C. How much potassium nitrate will precipitate out if the solution is cooled to 25°C?

13 **Suppose you dissolve 30 g of sodium chloride in 100 cm^3 of water at 100°C and boil away 50 cm^3 of the water.**
a) How many grams of sodium chloride will remain in solution?
b) How many grams will precipitate out of solution?

14 **Suppose you dissolve 40 g of potassium nitrate in 100 cm^3 of water at 100°C.**

a) If half the solution is poured out, how many grams of potassium nitrate will the remaining solution contain?
b) Now, instead of pouring out part of the solution, you boil away 50 cm^3 of water. How many grams of potassium nitrate will remain in solution at 100°C?
c) If the solution remaining in (b) were cooled to 25°C, how much potassium nitrate would precipitate out of solution?

EXPERIMENT FRACTIONAL CRYSTALLIZATION 5.6

In the preceding section we considered the separation of a mixture of sodium chloride and potassium nitrate. We predicted the results of a separation procedure based on the solubility curves in Fig. 5.4. The purpose of this experiment is to test the predictions by using the same quantities of solids and water as determined in the preceding section.

Before you begin the experiment there are a number of technical questions that you should consider. We chose to have the hot solution at 70°C. This choice of the temperature determined the volume of water needed to dissolve all of the potassium nitrate. Must you be sure when you filter the liquid that its temperature is indeed 70°C? Why or why not?

After you add 3.6 cm^3 of water to the 10 g mixture of solids (5.0 g of potassium nitrate and 5.0 g of sodium chloride), heat the mixture while stirring it and then filter.

A hot solution cools rapidly. To prevent this you can preheat the funnel, filter paper, and the test tube in which you will collect the filtrate by rinsing them with hot water.

What do you expect to have on the filter paper? What substance do you expect to precipitate from the filtrate?

When solids precipitate slowly out of solution, the crystals they form have characteristic shapes. What is the shape of the crystals that precipitated from the solution when it cooled?

The sodium chloride that you separated in your experiment was never dissolved. To observe its crystals you can prepare a saturated solution of it and allow it to evaporate overnight. How would you describe the shapes of the sodium chloride crystals? Is there another way, other than by observing the crystal shapes, that you could use to show that the two solids you separated are different?

5.7 EXPERIMENT PAPER CHROMATOGRAPHY

Hang a strip of filter paper streaked with ink in a graduate containing water, as shown in Fig. 5.6. When the color has risen up the paper to about 2 or 3 cm below the top, remove the paper and hang it up to dry.

How many different substances can you identify? Can you put the substances back together again and make black ink? Cut out each of the colored sections, and put each one in a separate test tube. Add between ½ and 1 cm^3 of water to each tube. Do the colored substances dissolve? Pour the liquids from all three test tubes together into one test tube. What color do you get?

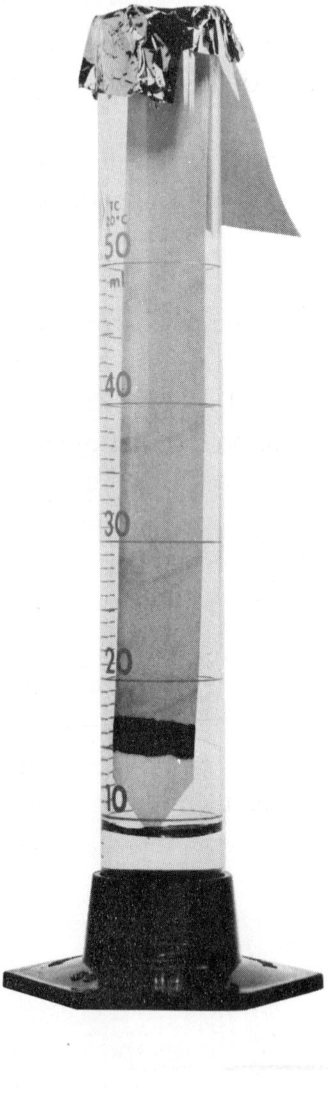

FIG. 5.6
One method of making a paper chromatograph of black ink. The point of the filter-paper strip extends into the water so that the ink streak is about 1 cm above the water. The strip is held in place by an aluminum-foil cover bent over the top. The walls of the graduate should be dry so that the paper will come in contact with water only from the bottom.

5.8 MIXTURES OF GASES: NITROGEN AND OXYGEN

In this chapter we have seen how we can separate mixtures of liquids, mixtures of solids and liquids, and mixtures of solids. We have not yet considered the separation of mixtures of gases. If we have a gas dissolved in a liquid, all we have to do is heat the mixture. (You did this in Sec. 4.12 when you heated a solution of ammonia gas in water.) When a glass of cold water warms up to room temperature, bubbles of air dissolved in the water appear, sticking to the sides of the glass. If we warm the water further, more air bubbles appear, and some rise to the top long before we reach the temperature at which water boils. Mixtures of gases alone, however, require different methods from those we have used in experiments so far.

Mixtures of gases are very common. Table 5.1 (page 71) lists four gases that are mixed together in the petroleum fraction called "fuel gas."

There are a number of ways of separating gases. One of them, which is widely used, is to cool the mixture until it condenses to form a liquid. Then we can make use of the different boiling points of the various liquids and fractionally distill the cold liquid. The gases are then collected one by one, as the boiling temperature levels off at new plateaus.

If we liquefy air and fractionally distill it in this way, we find that air separates mainly into two fractions: A glowing splint bursts into a bright flame when placed in one of them, but even a splint that is burning goes out when placed in the other. Neither gas turns limewater milky. The gas that causes the glowing splint to burst into flame is called "oxygen"; the one that does not is "nitrogen." These two gases together make up about 99 percent of the gases in air. Nitrogen makes up about 80 percent of the atmosphere, and oxygen about 20 percent. The densities, melting points, and boiling points of nitrogen and oxygen are given in Table 5.2.

Table 5.2

Gas	Density (g/cm^3)	Melting point (°C)	Boiling point (°C)
Nitrogen	1.2×10^{-3}	-210	-196
Oxygen	1.3×10^{-3}	-218	-183

The densities are given for atmospheric pressure and room temperature.

The cheapest way of obtaining oxygen and nitrogen is to condense air into a liquid and then fractionally distill it. Most of the oxygen and nitrogen commercially manufactured is produced by this method.

15 How could you separate ammonia gas from air?

5.9 LOW TEMPERATURES

The melting points and boiling points given in Table 5.2 are far below any temperature you can reach in your laboratory. How is it possible to cool something to such low temperatures?

One method of cooling gases depends on the fact that very highly compressed gases cool when allowed to expand. Figure 5.7 shows how this effect can be used to cool air to temperatures low enough to liquefy it. Air at very high pressure and room temperature flows down the long tube in the center and escapes through a small opening at the bottom. As it escapes, it expands and cools. When the flow is first started, the escaping air does not cool enough to condense into liquid air; but as this escaping cold air flows up past the long tube, it cools the air moving down inside the tube.

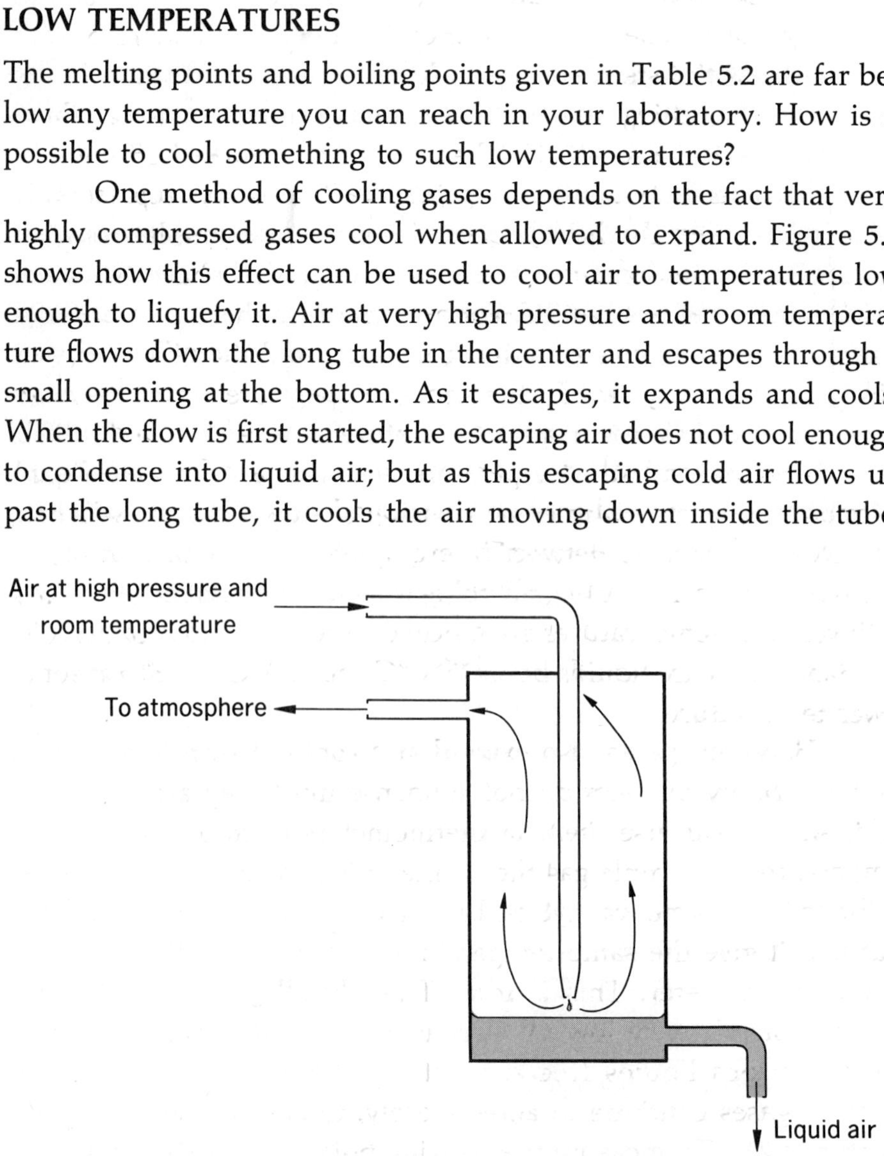

FIG. 5.7
A simplified diagram showing the essential steps in liquefying air.

Thus when the air inside escapes at the bottom, it is already cold and cools off still more on expansion. After the apparatus has run for some time, the expanding air cools enough so that some of it condenses into liquid and collects at the bottom of the apparatus. Of course, the whole apparatus shown in the figure must be well insulated to keep the inside cold. Actual liquid-air machines are more complicated than this simplified diagram shows, but many of them operate on this principle.

It is one thing to produce very low temperatures but another to measure them. The usual way of calibrating a liquid thermometer is to mark on the stem the liquid levels when the thermometer is placed first in melting ice and then in boiling water and to call these temperatures 0°C and 100°C. The scale is then marked off into 100 equal divisions between these points, each division representing 1°C. If we specify the liquid used—mercury, for example—we have then defined a temperature scale. Many liquids behave the same way between 0°C and 100°C. For example, two thermometers calibrated in this way, one containing mercury and the other toluene, will both read very nearly the same temperature when they are placed together in water at any temperature between 0°C and 100°C.

If we extend the temperature scale on a calibrated liquid thermometer by marking equal divisions below 0°C, we still find very close agreement between thermometers containing different liquids, such as mercury and toluene. This agreement continues until very low temperatures are reached, where we run into trouble. Substances that are liquids between 0°C and 100°C solidify at some lower temperature.

However, gases also expand and contract with temperature changes. Many of them do not condense until they are extremely cold, and we can use them in thermometers to measure very low temperatures. A simple gas thermometer is shown in Fig. 5.8. If we calibrate it the same way we calibrate a liquid thermometer, we find that it will give the same temperature readings as will a standard liquid thermometer. This is true of nearly all gas thermometers. Furthermore, at very low temperatures where thermometers containing various liquids freeze, most gas thermometers containing different gases continue to agree closely, thus extending our temperature scale. To measure the freezing point of oxygen, we need a gas that condenses below −218°C. Helium is such a gas and is used in gas thermometers.

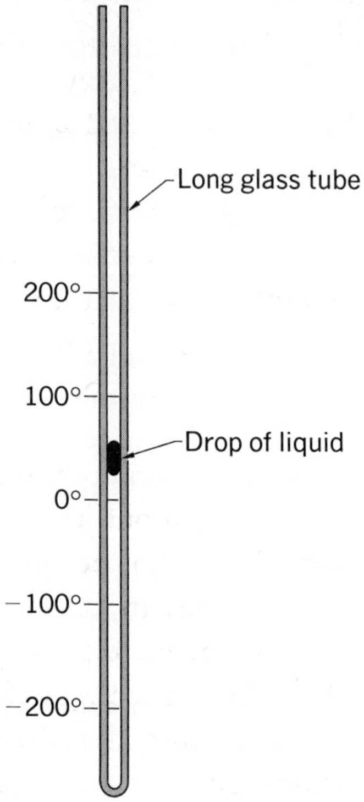

FIG. 5.8
A simple gas thermometer calibrated in degrees Celsius.

16† Liquid nitrogen will boil in a teakettle resting on a cake of ice. How do you account for this?

5.10 MIXTURES AND PURE SUBSTANCES

In this chapter we have seen that in learning how to determine the characteristic properties of substances, we have also found ways to separate different substances from each other. For example: A difference in density can be used to separate two solids; solids can also be separated by differences in solubility; a difference in the boiling points of different liquids enables us to separate them by fractional distillation.

Suppose we experiment with a piece of solid material to see if we can separate it into two or more substances. First we grind it up and mix it with water, stirring it thoroughly. We observe that some particles of a yellowish solid float on the surface while particles of a gray solid sink to the bottom. We skim off the floating material, whose density is obviously less than that of water. We dry it, call it fraction 1, and set it aside. Then we filter the water and the more dense solid that is in the bottom of the test tube. This solid, which remains on the filter paper, we dry, label fraction 2, and set aside also. We know that these two solids, fractions 1 and 2, are different substances because they have different densities.

We now test the filtrate to see if any material has dissolved in the water. Evaporating away the water, we find a small amount of white solid. This substance is different from both fractions 1 and 2 because it is soluble in water, and we call it fraction 3. The whole process of the separation of three fractions from a piece of solid material is diagrammed in Fig. 5.9.

We now have separated out three different substances, but perhaps each of these can be further separated. To find out, we use other separation methods. We may, for example, try to melt and even fractionally distill each of the fractions, or we may try dissolving them in different liquids. Suppose that boiling, melting, and mixing with alcohol and other liquids do not produce anything with characteristic properties different from those of the three fractions we have already found. By using all these various tools of separation again and again, we find that the characteristic properties of our three fractions remain unaltered. We have thus arrived at a collection of three substances whose properties are not changed by repeating any of these procedures. We call such substances "pure" substances.

Suppose we mix together all the pure substances that we obtain in this way and get back a material that has the same characteristic properties as the original sample. Then we say that the original sample was a *mixture* of the pure substances. For example, if you mixed all the fractions you separated in your experiment on fractional distillation (Expt. 5.1) you would get back a substance that has the same properties you started with. Similarly, mixing together the various fractions obtained from the distillation of liquid air would yield back ordinary air. Therefore, air is a mixture.

Note that many properties of a mixture are intermediate between the properties of the pure substances that form the mixture. For example, the density of air is between the density of nitrogen and that of oxygen. A mixture of alcohol and water will smell like alcohol; and if it contains enough alcohol, it will even burn.

However, it is not always possible to get back the original sample by mixing the parts separated by the various processes. For

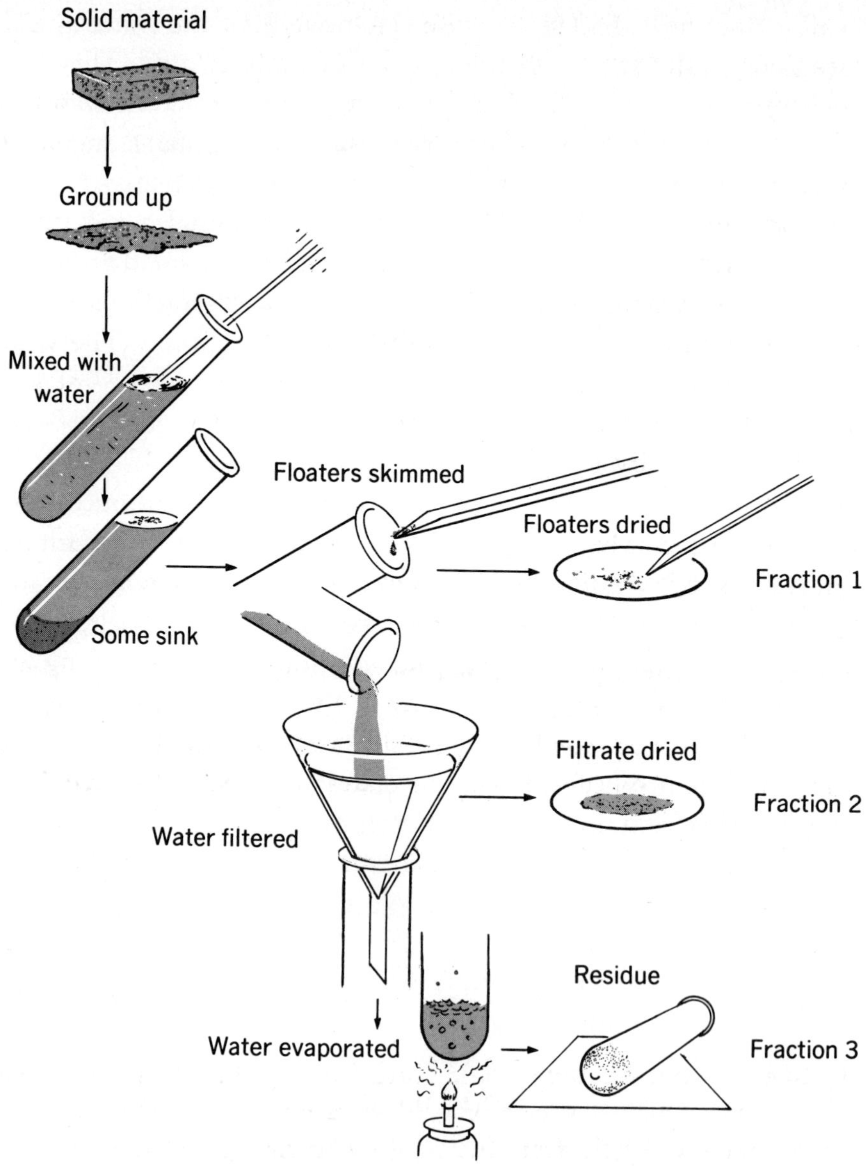

FIG. 5.9
A diagram of an example of the series of steps for separating three fractions obtained from a ground-up solid.

example, mixing the distillation products of petroleum will yield a liquid similar but not identical to the original sample. You know, of course, from your own experience that you cannot mix together the products of the dry distillation of wood and get back anything resembling wood. Wood is not, therefore, a mixture of the substances separated by distillation.

A simple example of a substance that breaks up when heated and is not a mixture is an orange powder called "mercuric oxide." If you try to determine its melting point, you will discover that when it is heated it gives off a gas that you can test and find to be oxygen. Furthermore, you can detect some droplets of a silvery liquid (mercury) in the test tube. Mixing the oxygen and the mercury together, however, will not give back the mercuric oxide. The two components remain separated as a gas and a silvery liquid.

Mercuric oxide is a pure substance that cannot be separated into simpler substances by most of the methods we used to separate mixtures. When it is separated by heating, it cannot be put back together simply by mixing.

---

17 Suppose you mixed together all the fractions you obtained from the fractional distillation of the liquid in Expt. 5.1. What do you think would be the properties of this liquid?

18 The substances you obtained by distilling wood (Expt. 1.1) when mixed together will not give anything like wood—even ground-up, finely powdered wood. What does this tell you about the substances in wood as compared with the substances you obtained in the distillation?

For Home, Desk, and Lab

---

19 In earlier times people would search for and find sandy stream beds in which small particles of gold were mixed with the sand. They separated this gold from the sand by "panning." Find out how this was done. What characteristic property of the substances made panning possible?

20 Figure 5.2 shows four oil wells drilled into oil-bearing porous rock. Can you suggest some method, without drilling deeper, for getting more oil from well *D* after the oil level drops below the end of the well?

21 a) In Table 5.1 (page 71), what fractions would be liquid at room temperature (20°C)? Which would be solids? Which would be gases?
b) You can see from the table that pentane is not an ingredient in any of the common products listed. How can you account for this?

22 Using the data in Table 5.1 (page 71), draw and label a possible distillation curve for a mixture of hexane, nonane, and tetradecane.

23 When ethanol is to be used for commercial or industrial purposes other than as a beverage, it is customarily "denatured"; that is, a small quantity of another substance is added to it so that it is unfit for use as a beverage. What, in general, do you think some of the properties of the added substance might be?

24 How would you separate a mixture of powdered sugar and powdered citric acid?

25 The mineral called "Gay-Lussite" appears to be a pure substance, but it is actually a mixture composed of calcium carbonate (limestone) and sodium carbonate (soda ash) and water. Describe how you would go about separating these three substances from the rock. Some properties of calcium carbonate and sodium carbonate are listed in the following table.

Property	Calcium carbonate	Sodium carbonate
Melting point	Decomposes at 825°C	851°C
Solubility in alcohol	Insoluble	Insoluble
Solubility in hydrochloric acid	Soluble	Soluble
Solubility in water	Insoluble	7 g/100 cm^3 at 0°C; 45 g/100 cm^3 at 100°C

26 If you have 100 cm^3 of water at 100°C, saturated in both potassium nitrate and sodium chloride, what happens if the temperature is lowered to 10°C? (Refer to Fig. 4.3, and assume that the solubility curves of these substances are the same as in the figure, even when they are dissolved together.)

27 You can use paper chromatography to separate the components in many common substances. Here are some you can work with at home: tomato paste, different colors and brands of ink, the coloring in leaves and vegetables (grind the leaves first in alcohol), and flower petals.

28 Chlorophyll can be extracted from leaves by grinding them with alcohol to give a dark-green solution. By careful application of paper chromatography, bands of yellow and red color, as well as green bands, can be detected. What other reason do you have to suspect the presence of substances producing these colors in leaves? Why don't you ordinarily see them?

29 As liquid air boils away, the remaining liquid becomes richer in one of the two gases—nitrogen or oxygen. Which one is it? How do you know?

30 a) How would you calibrate the simple gas thermometer shown in Fig. 5.8 to read in Celsius degrees?

b) Which end of the liquid drop would you take as a reference point?

31 What would you do to separate (a) alcohol from water, (b) sodium chloride from sodium nitrate, (c) nitrogen from oxygen?

32 A sample of a liquid was boiled for 12 min. During that time the boiling point remained constant and the volume was reduced to half. Is the liquid a pure substance?

33 Suppose you had a mixture of sand and salt in a small box.

a) How could you separate these substances?

b) How would you determine the ratio of the mass of sand to the mass of salt?

c) If you were mixing sand and salt together, what mass ratios would it be possible for you to make?

Compounds

6

At the end of the last chapter, we mentioned a pure substance, mercuric oxide, that can be broken down into two different pure substances—mercury and oxygen—by heating. The properties of these two components are quite different from those of mercuric oxide. Furthermore, the mercury and oxygen cannot be put back together to form mercuric oxide simply by mixing. In the next two sections, you will study two other examples of breaking up pure substances, using two different methods to do so.

6.1 EXPERIMENT DECOMPOSITION OF SODIUM CHLORATE

What happens when you place about 5 g of sodium chlorate in a test tube and heat it with two alcohol burners as shown in Fig. 6.1? (**CAUTION:** Be sure to use a glass-wool plug and a test tube that is clean and dry.) The apparatus is arranged so that you can collect several test tubes of any gas that is given off. When you have filled all the test tubes with gas, take the stopper out of the test tube you are heating so that no water will be accidentally sucked back into the hot tube. A glowing splint may help you to identify the gas.

To make sure that all the solid in the test tube has been heated thoroughly, continue to heat the material for about 10 min

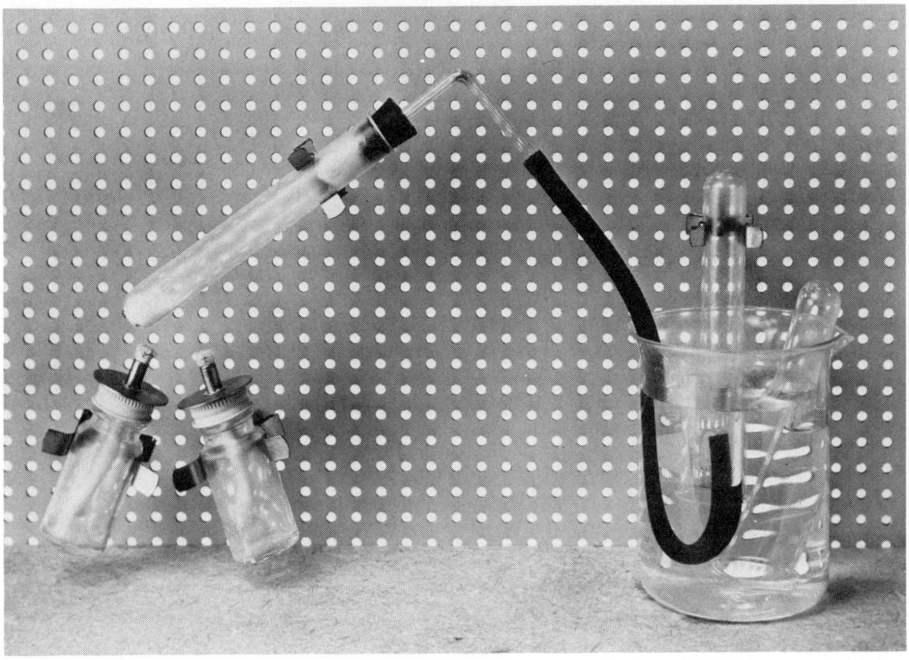

FIG. 6.1
Apparatus for decomposing sodium chlorate and collecting any gas that is given off. The glass-wool plug in the top of the test tube prevents the sodium chlorate from spattering up and coming into contact with the stopper.

after no further change is observed. You may want to rotate the tube to do this, but don't burn your fingers.

Does the material left in the test tube have the same melting point as the material you started with?

How does the solubility in water of the solid remaining in the test tube compare with the solubility of the sodium chlorate?

What do you conclude from this experiment about what happens when sodium chlorate is heated? Can you get back the sodium chlorate by mixing the gas and the remaining solid?

It is not possible from the tests you have made in this experiment to identify the substance left in the test tube. More experiments must be performed. One of these, the determination of the solubility of the substance as a function of temperature, has been done as follows.

A saturated solution (in water) of the substance remaining in the test tube was made at 100°C. Approximately 5 g of this hot liquid was poured into a previously massed evaporating dish. When the solution had cooled to 80°C, another portion of liquid was poured into a second evaporating dish. This process was continued, liquid being poured off at 60°C, 40°C, and 20°C.

The concentration of the saturated solution in each dish was determined by using the same procedure as in Expt. 4.1. Each dish of saturated solution was accurately massed. Then the solution was evaporated to dryness with an alcohol burner, and the mass of solid was determined.

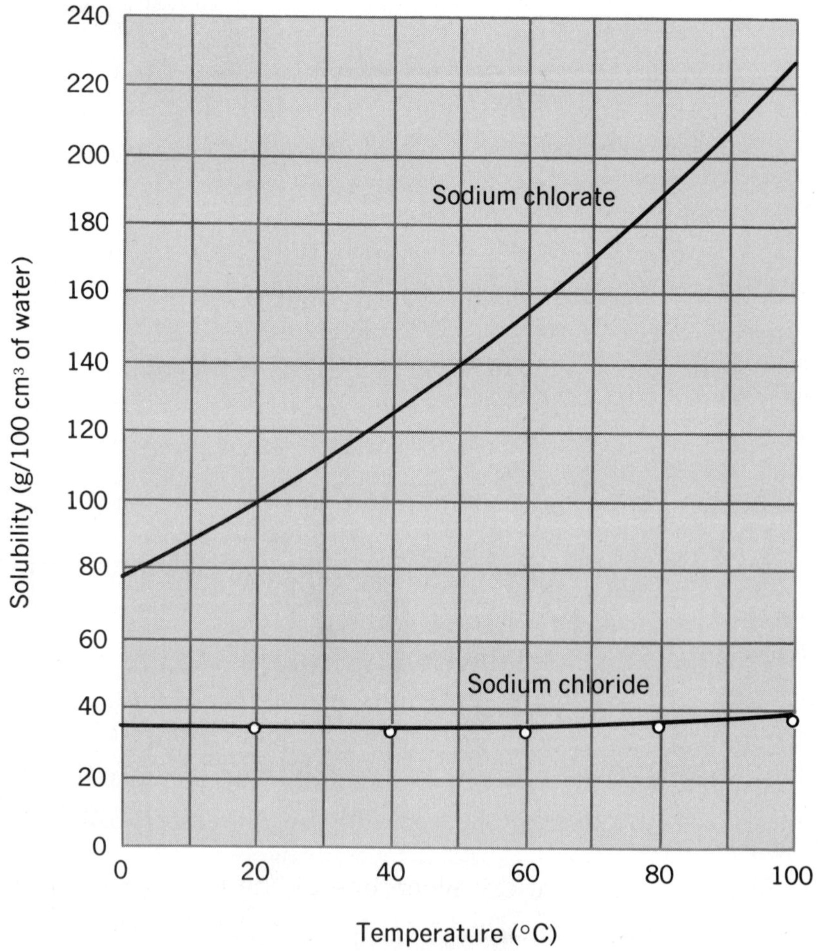

FIG. 6.2
The circles on the graph are data points obtained by measuring the solubility of the solid that remained after the sodium chlorate was decomposed. The line labeled sodium chloride is the solubility curve of sodium chloride shown in Fig. 4.3. The curve labeled sodium chlorate is included for comparison.

The mass of the water evaporated was calculated from the difference in the mass of the saturated solution and the solid. From these data the solubility of each sample was calculated. After converting the data to g/100 cm^3 of water, the solubilities were plotted in Fig. 6.2. Note that the data points lie close to the solubility curve for sodium chloride. They certainly do not fit the solubility curve for sodium chlorate, which also is shown on the graph. It thus seems very likely that the solid resulting from heating sodium chlorate is sodium chloride. Sodium chlorate, a pure substance, can be decomposed into two pure substances that have entirely different characteristic properties from those of the compound sodium chlorate.

1† Suppose that in decomposing sodium chlorate, you did not heat the sodium chlorate long enough to decompose all of it.

a) What substances would be left in the test tube?

b) How would you separate them?

EXPERIMENT DECOMPOSITION OF WATER 6.2

Water is one of the pure substances we have separated out of many mixtures. But we have been unable to separate water itself into other substances by the methods we have used in the previous chapter. After heating, distilling, and freezing water, we always end up with plain water. To separate it into simpler substances, we shall need a new method, different from any we have used so far to accomplish such a separation. Such a method, first used about 1800, makes use of electricity.

Set up the apparatus shown in Fig. 6.3. Connect the wires to the battery and see if anything happens. Since this reaction is very slow with pure water, it is necessary to add something to speed up the process. Try adding from 10 to 30 cm^3 of sodium carbonate solution to the water.

Disconnect the battery when one of the test tubes is nearly full of gas, and mark the volume of gas in each tube with a grease pencil or rubber band. You have probably heard that water is made up of hydrogen and oxygen. Test the gases to see if you come to the same conclusion. Be sure to remove the test tubes from the water in such a way that you do not lose the gases you wish to test.

Measure the volumes of the gases in the test tubes. To compare these two volumes, divide the volume of the hydrogen by the

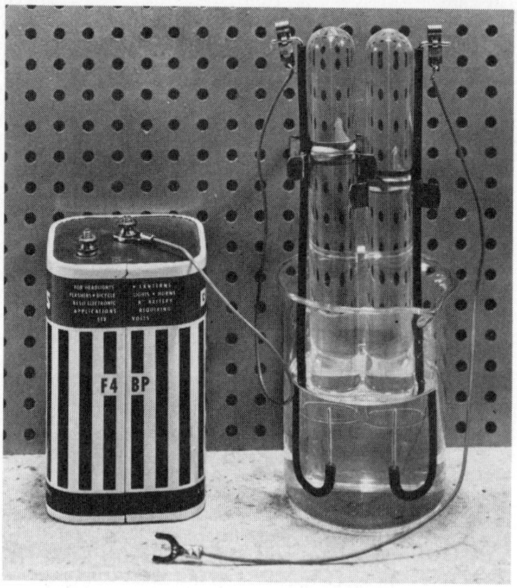

FIG. 6.3
Apparatus for decomposing water. The electrodes are clamped alongside the inverted test tubes and connected to a battery. Any gas that forms on either of the stainless-steel electrode tips will be collected when it rises and displaces water from its test tube. No reaction is observed until a small amount of sodium carbonate solution is added to the water.

volume of the oxygen. This gives you the ratio of the volume of hydrogen to the volume of oxygen. Compare your ratio with the ratios obtained by students who used different amounts of sodium carbonate solution mixed with the water. Does the amount of solution added to the water affect this ratio? Calculate the ratio of the mass of the hydrogen to the mass of the oxygen.

Repeat the experiment, this time collecting both gases together in one test tube and then igniting the mixture of gases. Be sure the two electrodes do not touch, or no gas will be produced. What do you observe?

——— ——— ———

A pure substance that can be broken up into two or more pure substances is called a *compound.* The characteristic properties of a compound are quite different from those of its components.

---

2 In decomposing water, suppose you had filled many test tubes with gas, adding water to the beaker as it disappeared but never adding more sodium carbonate. You would always have found the ratio of hydrogen to oxygen produced to be constant. What does this tell you about the source of the gases? About the sodium carbonate?

3† a) What is the total mass of oxygen and hydrogen that can be produced by the decomposition of 180 g of water by electrolysis?
b) If all the hydrogen produced were burned in the air to form water, what mass of water would result?

4† **Two test tubes contain equal volumes of gas at atmospheric pressure. If one contains oxygen and the other helium, is the mass of gas in both tubes the same?**

5 **From your data in Expt. 6.2, calculate the ratio of the mass of oxygen produced to the mass of hydrogen produced. How does your ratio compare with those of the other members of your class?**

THE SYNTHESIS OF WATER 6.3

In the last experiment you mixed oxygen and hydrogen gas together in a test tube. Nothing happened until you ignited the mixture. Before ignition, the gases could have been separated by cooling them down until the oxygen condensed at −183°C, leaving the hydrogen as a gas. But this method would not work after they were ignited. As a result of the ignition, the two gases combined to form a compound, water vapor, which cannot be separated into hydrogen or oxygen by condensation. The combining of substances to form a compound is called "synthesis." This process is just the opposite of decomposition.

When you electrolyzed water, you found the ratio of the volume of hydrogen to the volume of oxygen produced. No matter how much water you decomposed, the ratio remained the same. Of course, you would have expected this, since all the water you used came from the same source. But can we combine hydrogen and oxygen in any different proportions, or will they combine only in the same volume ratio?

To answer this question, we could try adding different volumes of hydrogen to a fixed volume of oxygen, igniting each mixture and measuring how much of each gas, if any, remains uncombined.

We have done this experiment. First we filled a special shock resistant glass tube with water and inverted it in a tray of water. A volume V of oxygen was then bubbled into the bottom of the tube, rising to the top as shown in the photograph at the left in Fig. 6.4(a). (The water levels are marked by rubber bands.)

We then added the same volume V of hydrogen to the tube (see the middle photograph in part (a) of Fig. 6.4). Next, we ignited the mixture of gases by an electric spark through the mixture. After

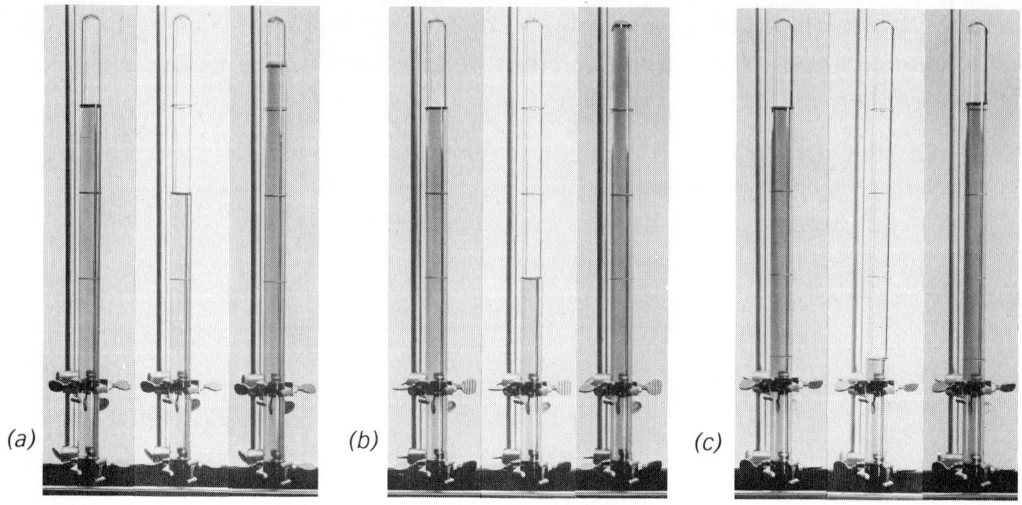

FIG. 6.4
The photograph on the left in (*a*) shows a glass tube, originally filled with water, after a volume V of oxygen was bubbled into the bottom of the tube. The middle photograph shows the same tube after the same volume V of hydrogen was added to the oxygen. The photograph on the right shows the tube after the mixture was ignited by an electric spark through the gas. Note that a considerable amount of gas remains unreacted.

In (*b*) we see the same experiment shown in part (*a*) except that the volume of hydrogen used was 2V. Note that all of both gases reacted except for a small bubble at the top. The volume of this bubble is no greater than the experimental error in filling the tube.

In (*c*) the volume of hydrogen used was 3V. As in (*a*), a considerable volume of gas remains unreacted.

ignition a considerable amount of gas remained unreacted, as shown by the photograph on the right in Fig. 6.4(a).

Fig. 6.4 (b) shows the same experiment repeated but this time the volume of hydrogen used was 2V. In this case practically all of the mixture reacted. (Only a small bubble of gas remained unreacted.)

In the experiment shown in Fig. 6.4 (c) the volume of hydrogen in the tube before ignition was 3V. An appreciable amount of gas did not react.

From Fig. 6.4 you can see that the ratios of the volumes of hydrogen to oxygen before ignition in (a), (b), and (c) were 1/1, 2/1, and 3/1 respectively. When tested after ignition, the gas remaining in (a) was oxygen; that in (c) was hydrogen. (The volume of gas remaining in (b) was too small to identify.) The increased amount of water in the tubes came from the tray of water. The water produced in the reactions amounted to only a few drops.

The results of this experiment are shown in Fig. 6.4 and Table 6.1. As you can see, the ratio of the volume of hydrogen to the volume of oxygen that combined to form water in each case was the same, regardless of the ratio of the volumes of the two gases in the mixture before they were ignited.

Table 6.1

Tube	Initial volume of oxygen (cm^3)	Initial volume of hydrogen (cm^3)	Final volume of gas (cm^3)	Volume of oxygen V_O that combined (cm^3)	Volume of hydrogen V_H that combined (cm^3)	Ratio V_H/V_O
1	25	25	12.4 oxygen	12.6	25	1.98
2	25	50	0.8	25	49.2	1.97
3	25	75	24.6 hydrogen	25	50.4	2.02

This is the ratio of hydrogen gas to oxygen gas obtained by the careful electrolysis of water in Expt. 6.2. The experiment in Fig. 6.4 shows that, over the range of volumes used, if the volume of either gas was greater than that needed for a ratio of 2, only part of the gas reacted. The excess remained uncombined.

6† If 18 g of water is decomposed into hydrogen and oxygen by electrolysis, 16 g of oxygen and 2 g of hydrogen are produced. Using the table of densities in Chap. 3 (page 43), find (a) the volume of water decomposed and (b) the volume of hydrogen produced.

7 You mix 100 cm^3 of oxygen with 200 cm^3 of hydrogen. The volumes of both gases are measured at atmospheric pressure and at room temperature.

a) Calculate the mass of oxygen used and the mass of hydrogen used.

b) If you ignited the mixture, what mass of water would result from the reaction?

c) What volume of water would be produced?

8 If in Question 7 you had used 100 cm^3 of oxygen but only 50 cm^3 of hydrogen, what mass of water would have resulted?

9† Three tubes are filled with a mixture of hydrogen and oxygen in a manner similar to that used in Fig. 6.4. If the three tubes contain the following volumes of hydrogen and oxygen, what is the volume of the ***unreacted*** gas remaining in each tube after they are ignited?

Tube	Volume of oxygen (cm^3)	Volume of hydrogen (cm^3)
I	25	75
II	50	50
III	25	50

6.4 EXPERIMENT SYNTHESIS OF ZINC CHLORIDE

You have seen that hydrogen and oxygen combine in a definite ratio, no matter how much of each we mix together and ignite. In that case, the reaction involves two gases. Let us now investigate what happens when we dissolve a metal in an acid: in this case, zinc in hydrochloric acid.

FIG. 6.5
Zinc is dissolved in 10 cm^3 of hydrochloric acid in a test tube, which is placed in a beaker of cold water to keep the solution from getting too hot.

In this experiment, everyone in your class will use the same amount of hydrochloric acid, but different groups will add different amounts of zinc. When the reaction is complete, you will determine the mass of zinc that reacted. You will then evaporate the remaining liquid and you will mass the solid residue. Then each group will calculate the ratio it found for the mass of zinc reacted to mass of solid product formed (its name is "zinc chloride").

The reaction between the zinc and the acid will produce much heat. To keep the mixture cool, you can perform the reaction in a large test tube placed in a beaker of cold water as shown in Fig. 6.5.

Mass out an exact amount of the zinc somewhere between 0.5 g and 4 g. Place the zinc in the test tube, and add 10 cm^3 of hydrochloric acid. What is the gas given off in the reaction? **Caution:** Be careful not to get any of the solution on your books or clothes. If some acid is spilled on your hands, wash them thoroughly with water.

The reaction at first is quite vigorous; but to make sure that it is complete, allow the mixture of zinc and acid to stand overnight. In the next period, pour the liquid from the test tube into an evaporating dish that you have already massed. If there is still zinc left over from the reaction, pour the solution so that the solid stays in the test tube. Wash the test tube, and any zinc remaining, with 5 cm^3 of water. Why should you add the washing water to the evaporating dish? Dry the leftover zinc, and mass it. How much of the metal reacted with the hydrochloric acid?

The evaporating dish can be placed on a burner stand and heated using an alcohol burner as shown in Fig. 6.6. If the solution begins to spatter, move the flame gently back and forth.

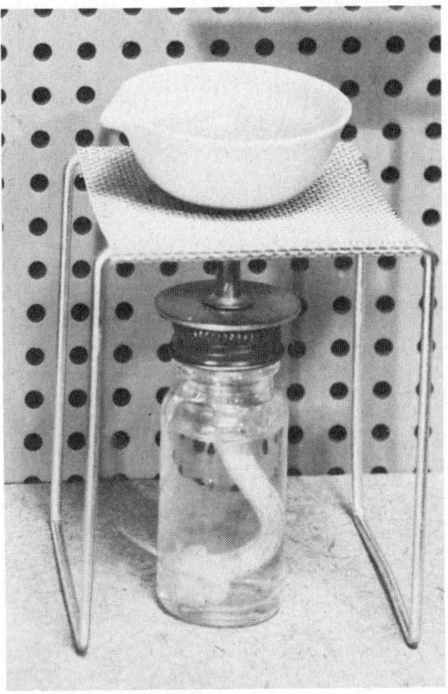

FIG. 6.6
Evaporating a solution in an evaporating dish heated directly over an alcohol burner.

Heat the material until it appears to be dry. Continue heating until the solid starts to melt and a tiny pool of liquid begins to form in the bottom of the evaporating dish. You can mass the dish and its contents as soon as they have cooled.

What is the ratio of the mass of zinc reacted to the mass of zinc chloride formed? Compare your results with those of your classmates. Did an excess of either zinc or hydrochloric acid affect your results? If the zinc chloride was not completely dry when massed, how did this affect your ratio?

---

10 **Suppose that in the synthesis of zinc chloride you dissolved 5 g of zinc.**

a) How much product would you get?

b) What would be the ratio of zinc to the product?

c) What would the ratio be if you dissolved 50 g of zinc?

11† In a certain package of seed corn the number of red seeds was 36, and the number of yellow seeds was 24. In a second package the number of red seeds was 51, and the number of yellow seeds was 34.

a) What is the ratio of the number of red seeds to the number of yellow seeds in each package?

b) What is the ratio of the number of red seeds to the total number of seeds in each package?

12 **Suppose that there are 10 boys and 15 girls in a class. (a) What is the ratio of boys to girls in the class? (b) What is the ratio of boys to total number of students in the class? (c) What would be the ratio of boys to girls if the class were three times larger but the ratio of boys to total number of students was the same?**

13† When various amounts of zinc react with hydrochloric acid, zinc chloride and hydrogen are produced. Which of the following ratios of masses between the various products and reacting substances are constant regardless of the amounts of zinc and acid mixed together?

$$(1)\ \frac{\text{Zinc added}}{\text{Zinc chloride produced}} \qquad (4)\ \frac{\text{Zinc used up}}{\text{Zinc chloride produced}}$$

$$(2)\ \frac{\text{Zinc used up}}{\text{Hydrochloric acid used up}} \qquad (5)\ \frac{\text{Zinc added}}{\text{Hydrochloric acid added}}$$

$$(3)\ \frac{\text{Zinc used up}}{\text{Hydrogen produced}} \qquad (6)\ \frac{\text{Zinc chloride produced}}{\text{Zinc used up}}$$

$$(7)\ \frac{\text{Hydrochloric acid used up}}{\text{Hydrogen produced}}$$

THE LAW OF CONSTANT PROPORTIONS 6.5

In the last two sections we studied the synthesis of two compounds: water from oxygen and hydrogen and zinc chloride from zinc and hydrochloric acid. We found that hydrogen and oxygen combine only in a definite mass ratio of 0.13. It does not matter in what proportion we mix these gases. When we ignite them, the hydrogen and the oxygen that react do so in a definite proportion to produce water. The zinc chloride that was produced in the last experiment was a result of the zinc's combining with the chlorine in the hydrochloric acid. The ratio of the mass of zinc that reacted to the mass of zinc chloride was constant. It was independent of whether you had an excess of hydrochloric acid or an excess of zinc. This means that the ratio of zinc to the chlorine with which it combined also was constant. We consider both water and zinc chloride to be compounds and not mixtures because each has characteristic properties quite different from those of the substances from which it is made. Do all substances combine in a constant proportion when they form compounds?

When you investigated the law of conservation of mass, there were two experiments in which compounds were formed. In one of these (Expt. 2.12), you heated copper and sulfur to make a new substance. Suppose we repeat the experiment, keeping the mass of copper constant and varying the mass of sulfur. Will we find that the ratio of the mass of copper that reacts to the mass of the product remains constant, independent of how much sulfur we use? This experiment has been done many times. As long as there is more than enough copper to react with all the sulfur, the ratio of the mass of the copper that reacted to the mass of the product remains fixed. But when there is an excess of sulfur, the ratio decreases. In this case, it appears at first glance that when copper and sulfur combine to form a compound, the ratio of the masses that react can vary.

The early chemists were in violent disagreement about the relative amounts of substances that react to form compounds. On one side was a distinguished French chemist, Berthollet (1748–1822). He claimed, on the basis of experiments like the one with copper and sulfur, that a pair of substances can combine in any proportion to form a compound. On the other side was another distinguished French chemist, Proust (1754–1826). He based his answer on evidence obtained from experiments that showed constant proportion, like the synthesis of water and of zinc chloride. Proust suggested a new law of nature, the law of constant proportions, which he stated in 1799: "We must recognize an invisible hand which holds the balance in the formation of compounds. A compound is a substance to which Nature assigns fixed ratios; it is, in short, a being which Nature never creates other than balance in hand." In plainer language, the law that Proust formulated can be stated as follows: When two substances combine to form a compound, they combine in a constant proportion. The ratio of the masses that react remains constant, no matter in what proportions they are mixed. If there is too much of one of the substances in the mixture, some of it will not react and will be left unchanged.

When the law of constant proportions was formulated, the evidence in its favor was much weaker than the evidence you gathered for the law of the conservation of mass at the end of Chap. 2 of this course. In spite of much evidence that supported Berthollet's stand and that Proust could not explain, Proust was confident enough to claim it as a law of nature.

The explanation of results like those in the copper-sulfur reaction that supported Berthollet's stand will be taken up later in the course.

---

14 If you make a solution of salt and water, over what range of values can you vary the mass ratio of salt to water at a given temperature?

15† a) Do you think that gasoline is a single compound? See Table 5.1.
b) Would you expect gasoline from different pumps to be the same?

6.6 EXPERIMENT A REACTION WITH COPPER

Some substances react very fast. As you have seen in Expt. 6.2, when a test tube of hydrogen and oxygen is ignited, the reaction is very fast indeed; it is explosive and is all over in a fraction of a second. Solids usually do not react so fast as gases. In this experiment, you will investigate how the reaction of finely divided copper with oxygen in the air proceeds with time.

Mass a crucible and add about 1 g of copper dust. Now find, as accurately as possible, the total mass of the copper and the crucible.

FIG. 6.7
Powdered copper in a crucible supported by a triangle over an alcohol burner. The wires of the triangle are bent so that they can be hooked into holes in the pegboard for rigid support.

Heat the copper, as shown in Fig. 6.7, for 2 min. While heating, watch the copper carefully. Does a reaction take place?

When the crucible is cool, find the mass of the crucible and its contents. Did the crucible and its contents gain or lose mass? Is your answer to this question evidence that a reaction has taken place? What do you predict will happen if you continue to heat the crucible for an additional 10–15 minutes? Try it.

Break up the contents of the crucible with a scoopula and examine the pieces. Do you think that all the copper has reacted? If not, what fraction do you estimate did not react?

Do not discard the black solid in your crucible. Pour the pieces of solid into a test tube and save them for use in the next experiment.

6.7 EXPERIMENT THE SEPARATION OF COPPER OXIDE FROM A MIXTURE OF COPPER OXIDE AND COPPER

If all the copper that you started with in the preceding experiment reacted with oxygen, then the black solid present at the end of the experiment would be the pure substance, copper oxide. If, however, not all the copper reacted, then the black solid would be a mixture of copper and copper oxide. In order to determine if the black solid is a mixture, we shall attempt to separate it.

A good way to separate a mixture of copper oxide and copper is to place the mixture in a solvent that will dissolve one of them but not the other. Copper will not easily dissolve in hydrochloric acid, as your teacher can show you. Copper oxide, on the other hand, is soluble in hydrochloric acid.

Place all the black solid you obtained in the previous experiment in a test tube. Add 5 cm^3 of hydrochloric acid and stir gently with a glass stirring rod for 5–10 minutes. (**CAUTION:** Be careful not to get any hydrochloric acid on yourself, books, or clothes. If some acid spills on your hands, wash them thoroughly with water.) After letting the solid settle to the bottom, slowly pour off the liquid into another test tube. Now wash the remaining solid several times with water.

What does the remaining solid look like? Was the black substance a mixture?

Do not discard the acid solution. You will need it for the next experiment.

6.8 COMPLETE AND INCOMPLETE REACTIONS

There are many reactions, like the reaction of copper with oxygen, that are slow, and it is difficult to tell when all of one of the reacting substances has been used up. Because the copper in your crucible changed to a black solid, you may have assumed that all the copper that was originally present in your crucible had reacted. This would have been an incorrect assumption, as the presence of copper in the black substance has shown.

Reactions like the reaction of copper with oxygen are called incomplete reactions because neither one of the reacting substances, copper or oxygen, is completely used up during the reaction. There are other reactions like the reaction of zinc with hydrochloric acid, which you studied in Sec. 6.4, that are complete reactions. Zinc will keep on reacting with hydrochloric acid until either one of the two has been completely used up.

Part of the controversy that existed among early chemists about the validity of the law of constant proportions (Section 6.5) could have been avoided if they had better analyzed what actually happened during their experiments. Some of their disagreements occurred because they did not understand the difference between complete and incomplete reactions. Specifically, they often mistook the masses they mixed for the masses that reacted. As your own experience shows, this mistake is easy to make.

For Home, Desk, and Lab

16 a) How would you have changed the procedure you followed in Expt. 6.1 if you had wanted to measure the density of the gas?
b) Would you have had to decompose all the sodium chlorate in order to determine the density of the gas?

17 Ten grams of sodium chlorate was strongly heated. After the heating it was found that 5 cm^3 of water at 100°C was required to completely dissolve a 2-g sample of the residue.
a) What was the solubility of the residue in grams per 100 cm^3 of water at 100°C?
b) Was any sodium chlorate left in the residue?

18 Suppose that the apparatus used in Expt. 6.2, Decomposition of Water, contained 100 cm^3 of water. Using this apparatus, a student collected 57 cm^3 of hydrogen and 28 cm^3 of oxygen. What fraction of the total volume of water was decomposed?

19 About a gram of salt is placed in a test tube half full of water and shaken; about a gram of citric acid is placed in a test tube filled with alcohol and shaken; some magnesium carbonate is dropped into half a test tube of dilute sulfuric acid; hydrochloric acid is poured into a dish containing magnesium. In each case, the solid disappears, and we say that it "dissolved." However, it is evident that two different kinds of dissolving have occurred. Divide the experiments into two classes. What did you observe that led you to divide them this way? What do you think you will observe in each case if you evaporate the solution to dryness?

20 A mass of 5.00 g of oxygen combines with 37.2 g of uranium to form uranium oxide.

a) How many grams of the oxide are formed?

b) What is the ratio of uranium to oxygen in this compound?

c) How much oxygen is needed to completely oxidize 100 g of uranium?

21 Discuss the process of boiling an egg in terms of complete and incomplete reactions. What may affect the time required for a complete reaction?

22 What evidence do you have for the following statements?

a) Zinc chloride is a pure substance and not a mixture.

b) Sodium chlorate is a pure substance and not a mixture of sodium chloride and oxygen.

c) Water is a pure substance.

23 A student heated a crucible containing 2.15 g of powdered copper until the mass of the contents of the crucible became 2.42 g. The black solid in the crucible was placed in hydrochloric acid, and 1.08 g of copper remained undissolved in the acid. What is the reacting ratio by mass of copper with oxygen?

24 Hydrochloric acid is a solution of the compound hydrogen chloride in water. Suppose there were two bottles containing hydrochloric acid in your lab. How would you determine whether the concentration of hydrogen chloride in each bottle is the same?

Elements

7

EXPERIMENT ZINC AND COPPER 7.1

When you heated a sample of finely divided copper in Expt. 6.6, its mass increased. This fact indicated that something was being added to the copper. A new substance, copper oxide, was formed. In Expt. 6.7 you dissolved the copper oxide in hydrochloric acid, and from the black powder you produced a blue-green solution.

What happened to the copper in all these reactions? Did it disappear forever, or is it still in the solution in some form? Putting a piece of zinc in the solution will provide you with an answer. **CAUTION:** As you know from the synthesis of zinc chloride, zinc in hydrochloric acid will generate heat. Therefore, for safety reasons, it is advisable to put the test tube containing the solution in a beaker half filled with water.

Put a piece of zinc in the solution you saved from Expt. 6.7 and observe the reaction. While the reaction is going on, break up the solid from time to time with a stirring rod.

After the zinc has finished reacting, pour off the solution from the remaining solid and wash the solid several times with water to remove the acid from it. The washed solid can be removed from the test tube and dried quickly by pressing it between two layers of paper towel.

Press a scoopula firmly against the paper towel and move it across the dry solid. Can you identify the solid?

1 **What would happen if you heated the solid that you recovered in this experiment?**

2 **How do you expect the total mass of solid recovered in Expts. 6.7 and 7.1 to compare with the initial mass of copper you heated in Expt. 6.6?**

7.2 ELEMENTS

Heating copper in air is not the only way to produce a reaction with copper. There are many other reactions, and they all have one feature in common: The product of the reaction always has a larger mass than the mass of copper. Neither heating nor electrolysis will make something out of copper that has a smaller mass than the copper that we start with originally.

Contrast this behavior of copper with that of sodium chlorate (Expt. 6.1). When you heat sodium chlorate, oxygen is given off and the resulting sodium chloride has a smaller mass. We reason, therefore, that sodium chloride is a simpler substance than sodium chlorate.

In Expt. 6.2 we electrolyzed water, causing it to decompose into hydrogen and oxygen. In the same way molten sodium chloride can be decomposed by electrolysis into chlorine and sodium. The mass of either the chlorine alone or the sodium alone is less than the mass of the sodium chloride from which they were produced. Thus, sodium and chlorine are simpler substances than sodium chloride. However, just as with copper, none of the methods that can be used to break up other substances work for either sodium or chlorine. Pure substances that do not break up by heating, by electrolysis, reactions with acids, and by similar methods are called *elements.*

It is, of course, not necessary to try to break up the many thousands of pure substances we know in order to find out which of them are elements. In many cases, we know substances are compounds simply because they can be made by combining other substances. For example, we can "burn" sodium in an atmosphere of chlorine gas. The resulting white solid has the characteristic properties of table salt. Thus, table salt is not an element; it is the compound sodium chloride.

Early chemists using this reasoning were able to select as possible elements a relatively small number of substances from the list of pure substances they knew. One such list, proposed by the French chemist Lavoisier, is given in Table 7.1. Lavoisier gave the following explanation of his table of elements: "Since we have not hitherto discovered the means of separating them, they act with regard to us as simple substances, and we ought never to suppose them compounded until experiment and observation have proved them to be so."

Note that we define an element in a negative way. We call a substance an element if we fail to break it down by using a given set of tools, such as heating to a thousand or so degrees, passing an

Table 7.1 Lavoisier's List of the Elements (1789)

Lavoisier's name	Modern English name	Lavoisier's name	Modern English name
Lumière	Light*	*Étain*	Tin
Calorique	Heat*	*Fer*	Iron
Oxygène	Oxygen	*Manganèse*	Manganese
Azote	Nitrogen	*Mercure*	Mercury
Hydrogène	Hydrogen	*Molybdène*	Molybdenum
Soufre	Sulfur	*Nickel*	Nickel
Phosphore	Phosphorus	*Or*	Gold
Carbone	Carbon	*Platine*	Platinum
Radical muriatique†	—	*Plomb*	Lead
Radical fluorique†	—	*Tungstène*	Tungsten
Radical boracique†	—	*Zinc*	Zinc
Antimoine	Antimony	*Chaux*‡	Calcium oxide (lime)
Argent	Silver		
Arsenic	Arsenic	*Magnésie*‡	Magnesium oxide
Bismuth	Bismuth	*Baryte*‡	Barium oxide
Cobalt	Cobalt	*Alumine*‡	Aluminum oxide
Cuivre	Copper	*Silice*‡	Silicon dioxide (sand)

*Light and heat are no longer considered substances and are, therefore, omitted from our present-day list of elements.

†*Radical muriatique, radical fluorique,* and *radical boracique* are not on the present-day list of elements because Lavoisier only assumed that they existed in order to support his theory that chlorine (oxymuriatic acid gas) was a compound and not an element. No substances with properties such as Lavoisier ascribed to these three elements have ever been found.

‡The last five elements on Lavoisier's list are now known to be compounds.

Table 7.2 The 105 Known Elements

Element	Symbol	Element	Symbol	Element	Symbol
Actinium	Ac	Hafnium	Hf	Praseodymium	Pr
Aluminum	Al	Helium	He	Promethium	Pm
Americium	Am	Holmium	Ho	Protactinium	Pa
Antimony	Sb	Hydrogen	H	Radium	Ra
Argon	Ar	Indium	In	Radon	Rn
Arsenic	As	Iodine	I	Rhenium	Re
Astatine	At	Iridium	Ir	Rhodium	Rh
Barium	Ba	Iron	Fe	Rubidium	Rb
Berkelium	Bk	Krypton	Kr	Ruthenium	Ru
Beryllium	Be	Kurchatorium	Ku	Samarium	Sm
Bismuth	Bi	Lanthanum	La	Scandium	Sc
Boron	B	Lawrencium	Lw	Selenium	Se
Bromine	Br	Lead	Pb	Silicon	Si
Cadmium	Cd	Lithium	Li	Silver	Ag
Calcium	Ca	Lutetium	Lu	Sodium	Na
Californium	Cf	Magnesium	Mg	Strontium	Sr
Carbon	C	Manganese	Mn	Sulfur	S
Cerium	Ce	Mendelevium	Md	Tantalum	Ta
Cesium	Cs	Mercury	Hg	Technetium	Tc
Chlorine	Cl	Molybdenum	Mo	Tellurium	Te
Chromium	Cr	Neodymium	Nd	Terbium	Tb
Cobalt	Co	Neon	Ne	Thallium	Tl
Copper	Cu	Neptunium	Np	Thorium	Th
Curium	Cm	Nickel	Ni	Thulium	Tm
Dysprosium	Dy	Niobium	Nb	Tin	Sn
Einsteinium	Es	Nitrogen	N	Titanium	Ti
Erbium	Er	Nobelium	No	Tungsten	W
Europium	Eu	Osmium	Os	Uranium	U
Fermium	Fm	Oxygen	O	Vanadium	V
Fluorine	F	Palladium	Pd	Xenon	Xe
Francium	Fr	Phosphorus	P	Ytterbium	Yb
Gadolinium	Gd	Platinum	Pt	Yttrium	Y
Gallium	Ga	Plutonium	Pu	Zinc	Zn
Germanium	Ge	Polonium	Po	Zirconium	Zr
Gold	Au	Potassium	K	*	

*One recently discovered element has not yet been named.

electric current, treating with acid, and the like. This always leaves the possibility that some still untried tool will succeed. A comparison of Table 7.1 with Table 7.2, which lists the elements known today, demonstrates the point.

Do we believe that a hundred years from now a list of elements will be as different from the present one as the present one is from Lavoisier's? We certainly do not. The reason is that today we know that the substances we call elements have many properties in common that set them apart from compounds.

Table 7.2 lists 105 substances that we now know to be elements. Only about one-third of these are very common, and most of the great variety of natural and synthetic materials are made from this rather small collection of thirty-odd ingredients.

---

3 While on a class field trip a student found a shiny rock that appeared to be a metal. When she returned to the classroom, she heated the rock and found that it lost mass. Could this rock be an element? Explain your answer.

4 How do you know that water, zinc chloride, and sodium chlorate are not elements?

5 Hydrogen chloride—a gaseous pure substance—can be decomposed into two different gases, each of which acts like a pure substance. On the basis of this evidence alone:
a) Can hydrogen chloride be an element?
b) Can either of the other two gases be an element?
c) Can you be sure that any of the pure substances mentioned is an element?

6† Lavoisier's list of the elements (Table 7.1) included *magnésie*, *baryte*, *alumine*, and *silice*, all of which are now known to be compounds. What do you think are the elements in each of these compounds?

TWO SPECIAL CASES: LIME AND OXYMURIATIC ACID 7.3

Two substances on Lavoisier's list of the elements are of special interest because they illustrate the type of reasoning he used in making up the list. One of these substances is lime, which can be produced by strongly heating common limestone. Lime was known to the Romans as early as 200 B.C. Many attempts were made to decompose lime: The substance was heated in air, heated in a vacuum, heated with carbon. Every attempt failed, and most people agreed that lime must be an element. There remained a few doubters, who thought that lime was probably a compound of a metal with oxygen. They thought that lime could not be decomposed simply because the usual agent, carbon, which worked with copper oxide and other oxides, was somehow not "powerful" enough to separate the oxygen from the metal.

As so often happens, new discoveries must wait for new experimental methods. In this case, the electric battery, invented by Alessandro Volta in 1800, was the necessary tool. In 1807 the English chemist Humphry Davy used an electric battery to try to decompose metal oxides. He had already used a battery to decompose molten potash and molten soda to obtain the new elements potassium and sodium. A similar procedure failed to work with lime because lime could not be melted. After much experimentation, however, Davy finally was able, by carefully electrolyzing moist lime, to produce tiny amounts of a new element he called "calcium."

Davy solved another interesting mystery just two years later. Whereas lime had for many years been wrongly thought to be an element, this time an element was believed to be a compound. A very reactive gas, slightly green in color, had been discovered in 1774 as a by-product of some experiments with muriatic acid. (Muriatic acid we now call "hydrochloric acid.") Various people experimented with the new gas and found it similar in many ways to the gases from other strong acids. Acids were considered in those days to be those substances that were strongly corrosive and had a sour taste. Since oxygen was believed at that time to be the active ingredient in all acids, this new substance was named "oxymuriatic acid gas." This name stuck for more than thirty years, mainly because it had been suggested by some very important and respected chemists. Several people tried, unsuccessfully, to decompose the gas. They always blamed their failures on poor methods and poor tools. Finally, in 1810 Humphry Davy, after working with his usual patience and brilliance for a full two years, announced the results of his long series of experiments. These experiments, in his own words, ". . . incline me to believe that the body improperly called oxymuriatic acid gas has not as yet been decompounded; but that it is a peculiar substance, elementary as far as our knowledge extends, and similar in many of its properties to oxygen gas. . . ." In order that people would no longer be misled into thinking it a compound, he suggested a new name, "based upon one of its obvious and characteristic properties—its color," and called it "chlorine." Davy's conclusion has stood the tests of both time and new techniques, and chlorine is included in the present-day list of elements.

EXPERIMENT FLAME TESTS OF SOME ELEMENTS 7.4

As you have seen, it is not easy to decompose some compounds. Nor is it easy to identify which of the 105 known elements they are made from by recognizing their characteristic properties, such as melting point, boiling point, and density. In fact, some elements exist in such minute quantities that we cannot obtain samples large enough to allow us to measure their characteristic properties easily. Fortunately, there is a characteristic property of elements that is often easy to measure, even for very small quantities. We shall investigate this property.

Place small quantities of different compounds, each of which contains sodium, on tiny loops of nichrome wire and hold them in a flame. Record the color you see for each of the compounds.

Next, try the same experiment first with copper and compounds that contain copper and then with strontium. Try a sample of a compound containing lithium and then a sample containing calcium.

How can you recognize sodium in a compound by such a flame test? How can you recognize copper? Can you distinguish among strontium, calcium, and lithium compounds by a flame test?

---

7 When you hold a small amount of sodium chloride in a flame, you observe that the flame is strongly colored yellow. What could you do to be sure that the color is due to the sodium and not to chlorine?

8† If you spill a few drops of soup or milk on a pale-blue gas flame when cooking, the flame changes to a mixture of colors with yellow being the most intense color. How do you explain this fact?

EXPERIMENT SPECTRA OF SOME ELEMENTS 7.5

In the preceding experiment, you found that some elements are easy to distinguish by the colors their compounds give when heated in a flame. Many elements are not so easily identified. For such elements we need a way to separate the mixed colors so that we can detect slight differences in color that the eye cannot detect. The first step is to spread out the various colors in light in the way they are in a rainbow. This spread of colors is called a "spectrum." You have seen the rainbow spectrum of sunlight produced by water droplets, either in rain or from a water sprinkler. An easy way to produce a spectrum of the light from a common light bulb is to sight at the light across the surface of a phonograph record. Here it is the regularly spaced grooves on the record surface that produce the rainbow spectrum. A device called a "grating spectroscope" produces a spectrum in much the same way, but it allows one to make more exact observations of the spectrum produced.

Use a simple grating spectroscope to examine light from a straightline filament bulb and also from a fluorescent lamp. Do you notice any difference? Look at glowing mercury vapor through the spectroscope. You will see a number of narrow colored lines. Each of these lines is a separate view of the slit in the front of the spectroscope, formed in a different color and at a different position from the other lines. These are called "spectral lines." The arrangement of lines produced by a light source is called the spectrum of the source (plural: spectra).

7.6 SPECTRAL ANALYSIS

If you look through a spectroscope at a light containing all colors, you will see a complete rainbow pattern. If some color is missing, the eye will see a dark line at the position of the missing color. If some color is present in abundance, a bright line is seen at its proper place.

With a good instrument, the sodium spectrum looks like Fig. 7.1. This shows that the yellow of the sodium flame is not just any yellow. It is a very specific color indeed, which has its own special place in the spectrum. It is a yellow made by no other element. The presence of this particular pair of lines always means that sodium is present in the light source. Even if the yellow color is hidden from the unaided eye by many other colors, the spectroscope will show the presence of sodium.

Figure 7.2 illustrates spectra obtained from compounds of calcium, lithium, and strontium. Although all these elements give

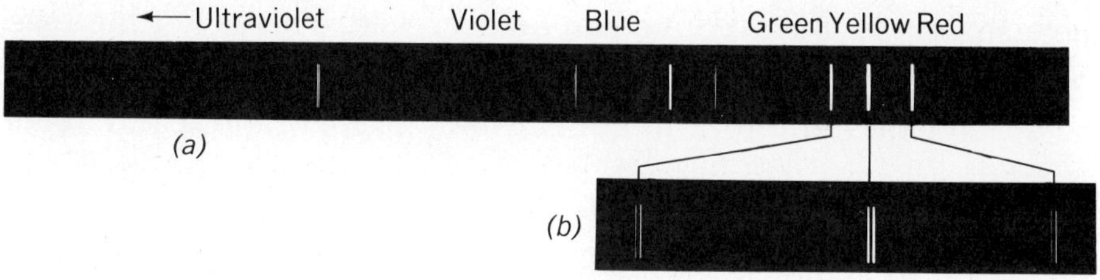

FIG. 7.1
(*a*) The spectrum of sodium. The line in the yellow is extremely intense. It has been photographed through an absorbing filter, because otherwise it could not be photographed well at the same time as the other lines. The close pair of almost invisible violet lines arises from a potassium impurity. (*b*) A part of the sodium spectrum has been photographed with a spectroscope that spreads out the light more, enabling us to see more detail. It shows that the yellow sodium line is really two lines very close together.

flame tests of nearly the same color, we see that each gives its own set of characteristic spectral lines when viewed through a spectroscope. The spectroscope thus enables us to distinguish one element from another.

We mentioned compounds of calcium, lithium, and strontium without specifying which compounds we were talking about. This may give you the impression that only the spectrum of one of

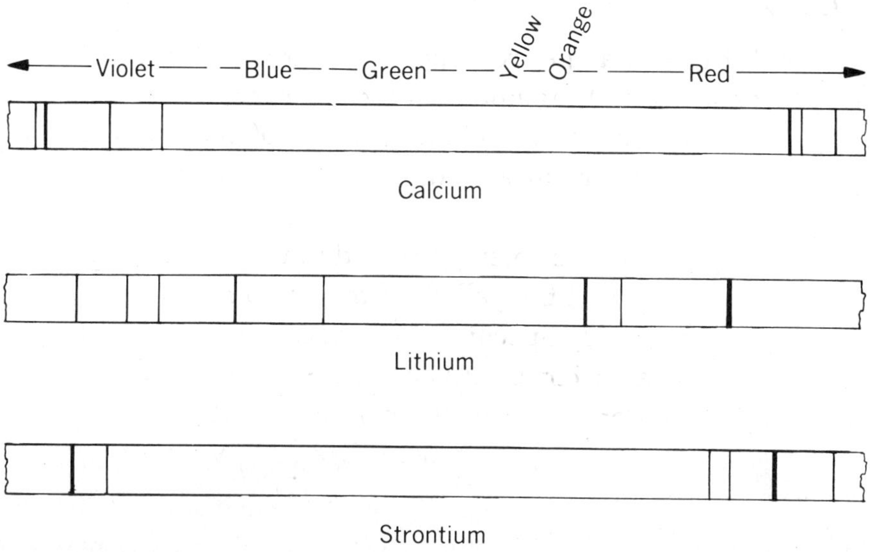

FIG. 7.2
The brightest lines in the spectra of calcium, lithium, and strontium.

the elements in a compound can be observed. This is not the case. While the flame of your alcohol burner is hot enough to produce the spectra of sodium, lithium, calcium, copper, and a few other elements, it is not hot enough to produce the spectra of other elements such as oxygen and chlorine. But if we heat a sample of a compound to a sufficiently high temperature (for example, by putting it in an electric arc) the spectra of all the elements in the compound will be observed. Under such conditions, the resulting spectrum is no longer simple. It will most likely contain complicated patterns of many closely spaced lines. Yet each element gives out its own spectrum, which is different from that of any other. It takes accurate measurements of the positions of spectral lines to identify an element. Once this has been done, the presence of that element has been definitely established.

Spectral analysis, first developed about 1860, is a powerful and delicate tool. It has made possible many important discoveries about the nature of matter. We know of almost a quarter of a million spectral lines that can be detected, and we know the exact position in the spectrum of more than 100,000 of them.

Spectral analysis, or spectroscopy, can be done on very small quantities of matter, such as a rare mineral or a very small sample of a biological material. Spectroscopy can even be used to determine the presence of different elements in distant objects like our sun and other stars.

Analysis of sunlight was one of the very early uses of the spectroscope in the study of unknown matter. Most of the spectral lines observed in sunlight could also be produced with known materials in the laboratory. However, during a solar eclipse in 1868, a new set of spectral lines was found in the spectrum of the light coming from the edge of the sun. This set of lines had never been seen before and could not be produced with any element known at the time. They were, therefore, thought to be from a new element which was given the name "helium," after the Greek word for sun. Eventually, the element was also detected on earth through the use of a spectroscope.

During the first few years of spectroscopy, five new elements were discovered that are present on earth in such small concentrations that they were previously unknown. For example, in analyzing the spectrum of the minerals found in the water of a certain spring in Germany, two lines of unknown origin were found in the blue region of the spectrum. This bit of evidence was enough to challenge Robert Bunsen, the German chemist, to search for a new element in the water. In order to isolate some of the pure element, which he named "cesium," it was necessary to evaporate 40,000 kg of spring water! In more recent times, spectral analysis has been one of the tools found helpful in identifying some of the new elements produced by nuclear reactions.

Time after time, this interplay between chemical analysis and spectral analysis has caused many complex substances to yield up the secret of their composition. Invariably, the results given by these two different methods agree completely.

We have now reached a high point in our study of matter. Through simple experiments you have been able to trace the main steps that led man to recognize how the material world is put together. It took hundreds of years but what was accomplished is really impressive: The millions of mixtures around us are made up from tens of thousands of compounds, which in turn are composed of only about a hundred elements. Compounds have characteristic properties that are different from those of the elements from which they are made. Yet the elements never really disappear. Their presence and their identity are revealed by their spectra, and they can always be extracted from their compounds. Elements seem to be permanent.

In science, however, it often happens that a new phenomenon is observed just when we think our understanding of a subject is complete. We find that our picture of the situation is incomplete; some important details need to be added.

In the case of elements, it was found that some substances that qualified as elements were in fact *not* permanent but changed on their own into other elements. The ground-breaking experiment leading to this discovery was made by the French physicist Henri Becquerel in 1896. We shall describe the essential points of his experiment in the next section.

---

9 Figure A (p. 118), shows some spectra observed with the same spectroscope that was used to obtain the spectra illustrated in Fig. 7.2. What elements can you identify in (*a*)? In (*b*)?

10 Suppose that you were given a sample of a substance. How would you try to find out if the substance were an element, a compound, or a mixture?

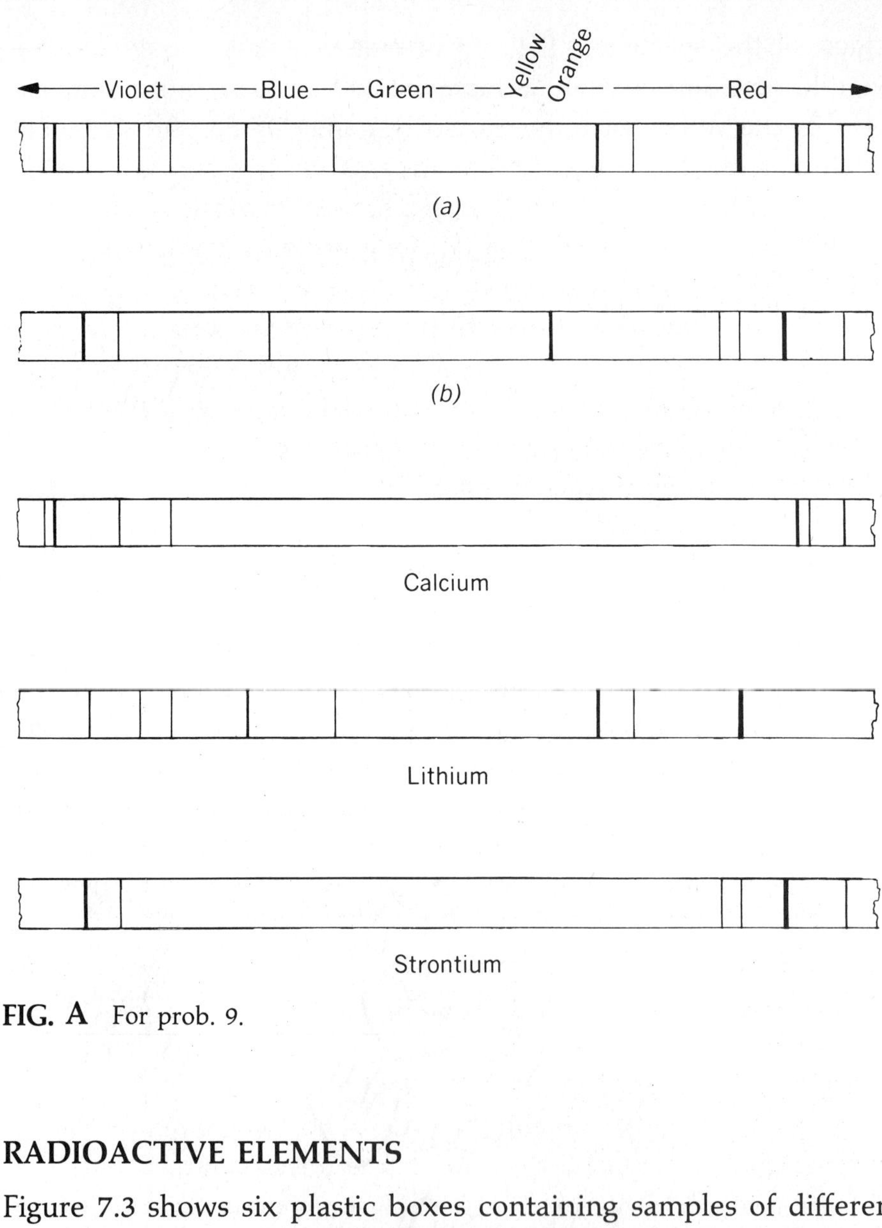

FIG. A For prob. 9.

7.7 RADIOACTIVE ELEMENTS

Figure 7.3 shows six plastic boxes containing samples of different materials placed on photographic film. The film was wrapped in black paper to shield it from the light. After three days the film was developed, and the result is shown in Fig. 7.4. Note the three white squares on the developed film where boxes A, C, and E were placed. Apparently, the material in these boxes gave off something that was able to get through the black paper and affect the film in the way that light does.*

*The details of this experiment are shown in the *Introductory Physical Science* film loop titled "Radioactive Substances I."

FIG. 7.3
Six different substances in small plastic boxes are placed on a photographic film enclosed in black paper. The boxes are left in position for three days before the film is developed.

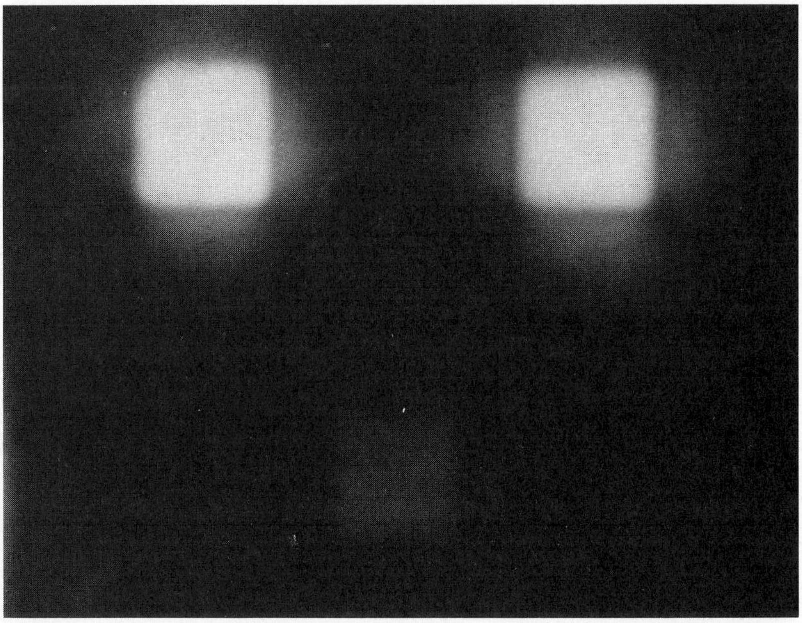

FIG. 7.4
The film in Fig. 7.3 after being developed. The three white squares appear where the boxes A, C, and E in Fig. 7.3 were placed.

The elements contained in the boxes (all in compounds, except box F) are listed in Table 7.3. Boxes A and B had two elements in common: sulfur and oxygen. However, box A, not box B, affected the film. This suggests that the uranium was the element that emitted or radiated something. This conclusion is supported by the fact that box C, which also contained uranium, left a mark on the photographic film.

Similarly, a comparison of the contents of boxes D and E suggests that the thorium in box E is the cause for the mark on the film.

Table 7.3

Box	Elements
A	Uranium, sulfur, oxygen
B	Sodium, sulfur, oxygen
C	Uranium, nitrogen, oxygen
D	Sodium, nitrogen, oxygen
E	Thorium, nitrogen, oxygen
F	Sulfur

Photographic film is not the only tool that distinguishes boxes A, C, and E from the others. If we place any one of these boxes near an instrument called a Geiger counter (Fig. 7.5), the counter will count or produce clicks that can be counted. Boxes B, D, and F do not affect the counter at all.

There are other elements listed in Table 7.2 that radiate something that affects a photographic film and a Geiger counter. These elements are called *radioactive* elements.

11† Elements *X*, *Y*, and *Z* form compounds *XY*, *XZ*, and *YZ*. Compounds *XY* and *YZ* are radioactive, but *XZ* is not. Which element is radioactive?

12† A piece of magnesium placed in hydrochloric acid causes hydrogen to be released. Evaporation of the resulting solution leaves behind a white solid. A similar reaction occurs when a piece of uranium is placed in hydrochloric acid. Would you expect this white solid to be radioactive?

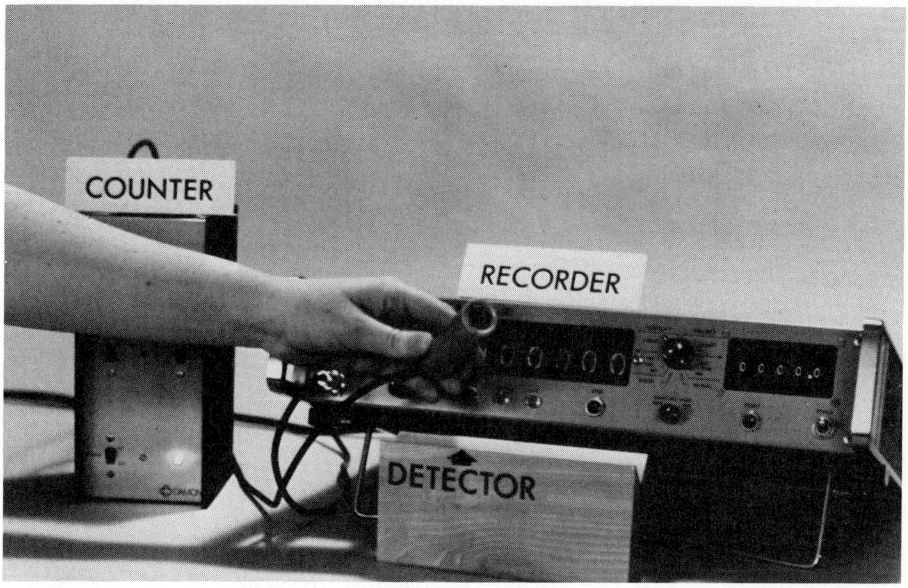

FIG. 7.5
A Geiger counter, used to measure the intensity of radiation from a radioactive source. The source is placed close to the end of the detector, which is held in front of the recorder. When a source is in place, numbers appear on the recorder that measure the radiation falling on the detector (from the *Introductory Physical Science* film loop, "Radioactive Substances I").

RADIOACTIVE DECOMPOSITION 7.8

What happens to a radioactive element when it radiates? This is a hard question to answer for the following reason. Some elements, such as uranium, radiate so weakly that it is extremely difficult to discover what happened to them. Other radioactive elements, such as polonium, emit such intense radiation that it is a simple task to determine what happened. However, because of its intense radiation, even a milligram (10^{-3} g) of polonium is a health hazard unless special precautions are taken. Despite the difficulties, what happens to a radioactive element has been studied for both weakly and strongly radiating elements.

In the case of polonium, spectral analysis reveals the secret. A freshly prepared sample of polonium has a spectrum that is characteristic of polonium. However, if we seal a sample of polonium in an evacuated glass tube and examine its spectrum a few weeks later, the spectrum will be quite different. In addition to the lines of polonium, the lines of helium and lead—two elements that were not present before—can be detected. If we check the spectrum after an additional month or two we find that the polonium lines get weaker, and the lead and helium lines get stronger.

Apparently, the radiation from polonium is connected with the change of polonium into lead and helium. Spectral analysis shows that other radioactive elements also change into different elements. In many cases helium is one of the elements that is produced. We refer to this process as *radioactive decomposition.*

Is radioactive decomposition different from the decomposition of water or other compounds? You remember that the products of the decomposition of water could be recombined to form the original compound. Lime, which Lavoisier thought to be an element, was finally decomposed into calcium and oxygen by electrolysis, but the calcium and oxygen easily recombine. It has been found possible to recombine all the elements that have been obtained from compounds. However, we have not been able to recombine the products of radioactive decomposition by the methods we have used to recombine elements into compounds.

What happens to a radioactive element when it is heated? Does heat change the rate at which it emits radiation? In many of the experiments you have done and read about in this course, temperature has had an important effect on what happened and how fast it happened. Wood decomposed when you heated it in a closed tube; the hotter the wood became, the faster it decomposed. Hydrogen does not burn by itself in air, but when it is heated with the flame of a burning match, it catches fire very easily, sometimes explosively.

To find out whether temperature has any effect on the intensity of radiation from radioactive substances, samples of these substances have been heated to very high temperatures, and they have been cooled to very low temperatures in liquid air. But it was found that temperature changes do not affect the radiation from a radioactive substance.

To sum up, the radioactive elements have all the properties of elements that you have studied in the preceding chapters. They form compounds with constant composition and have their characteristic densities, melting and boiling points, and spectra. They cannot be decomposed by ordinary heat, electricity, reaction with acids, and the like. They differ from nonradioactive elements in that they all affect a photographic film and all decompose into other elements. The rate at which they decompose cannot be changed by any of the means that affect the rate at which compounds decompose. That is why we call these substances elements; but to set them apart, we call them radioactive elements.

13† A single sample of uranium nitrate was left on a piece of photographic film for a week. During the week the sample was moved twice to new spots on the film. When the film was developed, the three "exposed" areas were found to vary in intensity. What can you conclude about (a) the temperature variation during the week and (b) the length of time the sample was in each position?

14 What are the two most important differences between the following two reactions?

a) Water → hydrogen + oxygen

b) Polonium → lead + helium

15 A radioactive sample at 20°C is placed near a device that counts the radiation coming from the sample. The counter records 1.0×10^2 counts/min. The temperature of the sample is then raised to 100°C. What does the counter then record?

A CLOSER LOOK AT RADIOACTIVITY 7.9

We have seen the evidence that some elements come from other elements. Now let us try to get a closer look at how this happens. Consider first the experiment with the photographic film described in Sec. 7.7. The light patches in Fig. 7.4 appear fairly uniform except for fading at the edges. However, the two patches at the top are more intense than the one at the bottom. How do these differences come about? Does the whole patch turn white or gray or dark gray?

Figure 7.6 shows an arrangement that was used to expose a photographic plate to a very small sample of polonium. Three plates were exposed for 66, 94, and 144 hours, respectively. Then the plates were developed, and the negatives were examined. The first negative showed a very light gray spot. The second showed a darker gray spot. And the third an even darker spot. The center of each of these spots was photographed through a high-powered microscope with a magnification of about 10^3 times. Figure 7.7 shows that under the microscope all three gray spots were seen to be made up of tiny black dots. Examine these photographs carefully and note that the black dots are about equal in size and blackness. Why, then, do the spots seem to be of different shades of gray when looked at without

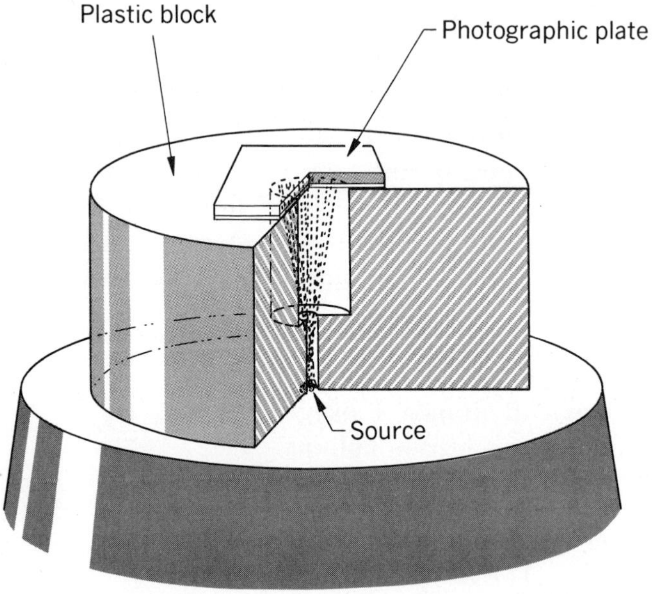

FIG. 7.6
A photographic plate exposed to a radioactive source of polonium.

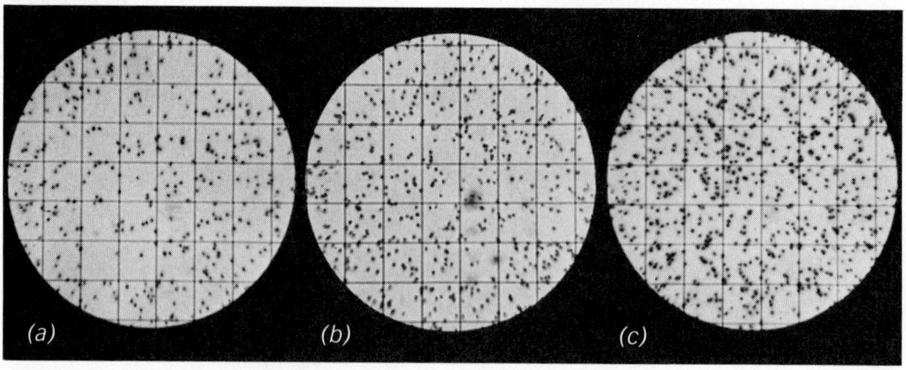

FIG. 7.7
Three photographic negatives obtained with the apparatus shown in Fig. 7.4. The negative shown at (*a*) resulted from the exposure of a photographic film to a tiny radioactive source for 66 hours. The one at (*b*) is the result of a 94-hour exposure to the same source; and the one at (*c*), of a 144-hour exposure. The negatives have been magnified 10^3 times. The three pictures show very small but equal-sized areas of the negatives.

the aid of a microscope? Even a quick glance will reveal that on the more exposed negatives there are on the average more black dots in each circle.

Table 7.4

Plate number	Exposure time (hours)	Number of dots on plate
1	66	369
2	94	474
3	144	724

The results of an actual count are shown in Table 7.4. The number of dots on the second plate is $474/369 = 1.3$ times the number on the first plate, and the time of exposure for the second plate was $94/66 = 1.4$ times as long as for the first plate. The third plate has 1.5 times as many dots as the second plate and was exposed 1.5 times as long. It appears that the number of dots depends directly on the exposure time. In other words, a plate exposed for 300 hours will show about three times as many dots as one exposed for 100 hours, and a plate exposed for 500 hours will show five times as many dots as one exposed for 100 hours. On the photographic film we see the result of something we have called radiation, which apparently comes from the radioactive source.

With a Geiger counter close to the polonium sample, we can hear the result of something that also appears to come from the polonium. It seems as though little invisible particles are actually flying off from the radioactive source and making the counter click when they hit it. If we bring the sample closer to the counter, the clicks do not get louder but only more frequent.

A simple device called a cloud chamber (Fig. 7.8) gives additional support to the idea that radioactive substances give off particles that can be counted. Figure 7.9 shows a photograph of a cloud chamber that contains a sample of polonium. You can see individual tracks coming from the sample.

The photographic film, the Geiger counter, and the cloud chamber all suggest that radioactive decompositions take place in discrete or countable steps. Combining this observation with the fact that elements are produced during radioactive decomposition

FIG. 7.8
A simple cloud chamber. The chamber itself is a cylindrical plastic box resting on a block of Dry Ice. A dark felt band, soaked in methanol, encircles the inside of the top of the chamber. A tiny radioactive source is placed on the end of the needle projecting into the chamber. The needle is supported by a cork stopper inserted into a hole in the wall of the cylindrical box. The bottom of the chamber is painted black so that the white fog tracks will be easily seen when viewed from above and illuminated from the side.

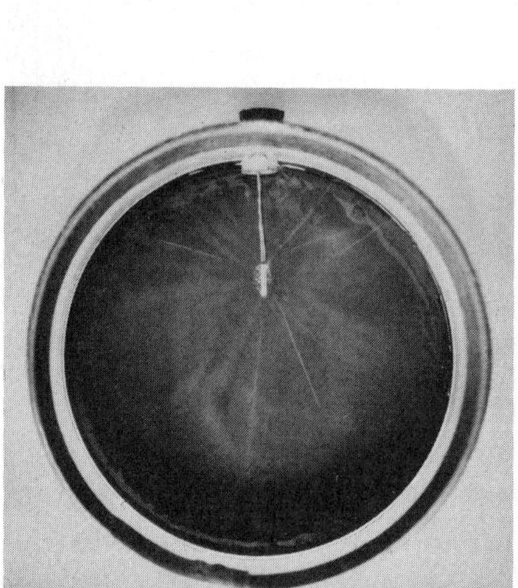

FIG. 7.9
Fog tracks produced in a cloud chamber by the radiation from a radioactive source placed on the end of a needle. How to prepare a cloud chamber and the formation of tracks are shown in the *Introductory Physical Science* film loop "Radioactive Substances II."

leads to a far-reaching conclusion: Elements are made up of discrete units, or some sort of tiny particles, called *atoms.*

Historically, the idea that elements are made up of atoms preceded the discovery of radioactivity. We have chosen to introduce this idea by discussing radioactivity because radioactivity enables us to count individual events in which new atoms are made (and old ones disappear).

16 **You look through a microscope at a photographic plate that has been exposed for 30 hours to a needle tip containing a tiny bit of radioactive substance. You count 563 dots. How many dots would you expect to count if the film were exposed for only 10 hours?**

17 **The number of black dots on the negatives in Fig. 7.7 is related to the exposure time. How do you suppose this relationship would work after very long exposure times?**

For Home, Desk, and Lab

18 When magnesium is put into hydrochloric acid, the metal reacts, a gas bubbles off, and a white solid is left behind after evaporation.
a) On the basis of this information alone, can you be sure which of the pure substances mentioned are compounds and which are elements?
b) We then find that the gas reacts just like hydrogen gas; it has the same characteristic properties. Also we find that hydrochloric acid can be decomposed into hydrogen and chlorine. Can the remaining solid be an element?
c) From the table of known elements (Table 7.2, page 110), which of the substances mentioned above are elements and which are compounds?

19 Use your grating spectroscope to examine the spectra of street lights, "neon" store lights, and any other light sources you can find (except the sun, which is too bright and will damage your eye if you look directly at it). Record your observations and compare them with those of your classmates.

20 a) If you heat a piece of blue vitriol (a blue solid), it loses mass and changes to a white powder. Which of the two substances might be an element?

b) If you dissolve the white powder in water and place an iron nail (iron is an element) in the solution, the nail will become coated with a thin layer of copper. What do you now conclude about the two substances in part (a)? Is either of them an element?

21 A plant absorbs various substances through its roots. Different elements of these substances concentrate in different parts of the plant. Suppose one of these elements is radioactive. How would you determine in what parts of the plant it concentrates?

22 In the film loop "Radioactive Substances I" the Geiger counter counted much faster when box A was placed next to it than when box E was. Is this what you would have expected on the basis of the brightness of the white patches in Fig. 7.4? Why or why not?

23 You saw in the film loop "Radioactive Substances I" that a Geiger counter counts something even if it is not placed near a radioactive source. Taking this fact into account, how would you measure the number of clicks per hour caused by a radioactive source?

24 The film loop "Radioactive Substances II" shows tracks being produced in a cloud chamber. From what you see in the loop, can you be sure that the tracks start at the source rather than end there? What does this tell you about the speed of the particles that leave the tracks?

The Atomic Model of Matter

8

A MODEL 8.1

When you look at a steel bar, pour water into a cup, or listen to the hiss of air leaking from a tire, you feel quite sure that matter is continuous. It does not look or act as though it is made of individual particles. But the black dots on the photographic film, the individual tracks in the cloud chamber, and the clearly countable clicks of the Geiger counter suggest a different picture—a picture in which matter looks grainy or discrete.

To help visualize what we mean by "grainy or discrete" in describing radioactivity, we can use a spray can spraying droplets of paint. Such a picture must not be taken too literally. Of course, we know that a piece of polonium does not look like a spray can, and we do not believe for a moment that the little particles that produce tracks in the cloud chamber are made of paint. We use a specific example or analogy as an illustration for a more abstract idea. The droplets of paint really stand for some kind of small particles, each of which can make a Geiger counter click or leave a track in the cloud chamber. If there are more particles, more tracks will be left; but the individual tracks are not affected by the number of particles. This abstract description of the radioactive process in terms of particles is an example of a "model," or a theory. This particle model clearly accounts for the increase in the blackening of a photographic plate with increasing exposure time.

But for a model, such as the particle model of matter, to be really useful, it has to offer us more than merely a convenient way of summarizing and accounting for facts we already know. A model must also enable us to make predictions. Here is an example of a very simple model and a prediction we can make from it: Suppose someone hands you a sealed tin can. You shake it and hear and feel something slosh around inside. From this simple experiment of shaking the can, you form a mental picture—a model—of what is inside. You conclude that the can contains a liquid. You have no idea what color the liquid is or what it tastes or smells like, but you feel sure that it has a property characteristic of liquids—it sloshes around inside a container. From this model (that the can contains something like a liquid) you can make a prediction: If you punch a small hole in the bottom of the can, liquid will drip out.

The simple model we have made for the behavior of the contents of the tin can was the result of just one experiment, shaking the can. It led to only one rather obvious prediction. Nothing about the model helped us to understand the liquid itself. We shall investigate a particle model of matter to see if it will help us understand more about the nature of matter. But first we shall do an experiment with a "black box," which is more complicated than is the tin can we have just described. The model you make to account

FIG. 8.1
A "black box."

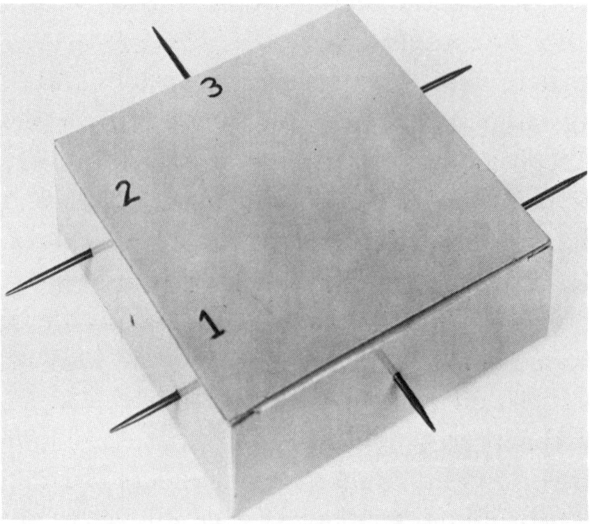

for the behavior of the black box and the predictions you make from the model will make it easier for you to understand a model for matter.

The box you will use is clearly man-made (Fig. 8.1). But you can experiment with it, organize the results of your experiment into a model, predict the results of new experiments. You can then test your predictions in ways quite similar to those we use when we investigate natural substances.

1† In addition to enabling us to summarize and account for the facts we have obtained from observations and experiment, what should a good model do for us?

2 You have two eggs in the refrigerator. One is raw and the other is hard boiled, but you do not know which is which. You spin one egg and it spins easily. You spin the other and it comes to rest after a very few rotations.

a) Suggest a model that will account for the different behavior of the two eggs.
b) On the basis of your model, what do you predict about the spinning of the two eggs if you boil both of them for 10 minutes?
c) Check your prediction at home.

EXPERIMENT A BLACK BOX 8.2

All the boxes you and your classmates will use in this experiment are the same. The first step is to find out as much as you can about these boxes without pulling the rods out of the boxes and, of course, without opening them. Look at one of them, shake it lightly, tilt it back and forth in various directions, and listen carefully to the sounds. You will find it very useful to write down your observations. This will help you to compare notes with your classmates so that you can arrive at a model, make predictions, and test them.

After you have done all the experiments you can think of, short of pulling out the rods (and, of course, opening the box), try to imagine in a general way what is inside the box that could account for your observations. This will be your model for the box. Do not be distracted by details. Do not, for example, try to name the objects inside the box; only describe them by the properties that you have found in your experiments. If you hear something sliding on one of the rods, you could equally well describe it as "a washer" or "a ring"; but the important point is that it is something with a hole in it through which the rod passes.

After you and your classmates have made models that account for your observations, predict what will happen as you pull out a particular rod. Also predict how this will affect the results of the tests you performed earlier. (Be sure to write down your predictions so that you can check them.) Then you or one of your classmates can remove this rod from only one of the boxes. Pulling out one of the rods may change things enough to prevent your checking your prediction of what would have happened had you pulled out another rod first. This is why only one box at a time should be used to test each prediction. If what happens confirms your prediction, you can use one of the other boxes to test your predictions about what would have happened if you had pulled out one of the other rods first. If, however, your first prediction was not confirmed, modify your model accordingly before further experimentation. Continue this process until you have arrived at a model in which you have confidence.

---

3 a) In investigating the black box you did certain things to it that you could easily undo. Give some examples.
b) In investigating the characteristic properties of a substance you can also do things that can easily be undone. Give some examples.
c) What did you do to the black box that you could not undo?
d) Does dissolving zinc in hydrochloric acid resemble any of the kinds of tests you mentioned in part (a) or part (c)?

8.3 THE ATOMIC MODEL OF MATTER

Let us now begin to build an atomic model for matter by reviewing some of the common properties of elements, compounds, and mixtures. From this we shall see that assumptions must be introduced into the model to account for these properties.

First, different samples of an element have the same characteristic properties. To account for this, we shall assume that an element is made up of only one kind of tiny particles, which we call "atoms." Different elements are made up of different kinds of atoms. Even with a high-powered microscope we cannot see atoms, and so they must be very small and there must be very many of them.

Second, in those experiments in which we worked with more than one element at a time, we found two quite different kinds of substances, which we labeled as compounds and mixtures. You will recall that, when elements react to form compounds, they combine only in a definite ratio. You saw evidence for this "law of constant proportions" in the burning of hydrogen with oxygen and in the synthesis of zinc chloride. Furthermore, the compound produced showed every evidence of being a new, pure substance. It had its own fixed set of characteristic properties, usually quite different from the properties of the reacting elements. The characteristic properties of carbon dioxide, for example, are entirely different from those of the carbon and oxygen that combine to make it.

Mixtures of elements, on the other hand, can be made in widely varying proportions of the basic ingredients and therefore do not show a law of constant proportions. Furthermore, the characteristic properties of a mixture can be varied over a wide range simply by varying the amounts of the elements being mixed together. And usually at least some of the characteristic properties of the individual elements (for example, color, solubility, smell, ability to react with other substances) are still present in the mixture.

We have clear evidence that the elements have not really disappeared when both compounds and mixtures are formed. In both cases, the pure elements can be freed and returned to their original form. The hydrogen and oxygen that were burned together to form water can be recovered by electrolysis of water. Nitrogen and oxygen can be mixed together to form air, and we can separate them into pure nitrogen and oxygen by liquefying the air and fractionally distilling it. Furthermore, even without separating the individual elements, we have some evidence that they are still present, even in a compound, by observing their spectra. In the compound sodium chloride, for example, we can clearly detect the spectral lines of the element sodium.

We can account for all this behavior by assuming that in each of these processes the atoms of the individual elements remain essentially unchanged. That is, neither their numbers nor their individual masses change. They only rearrange themselves. Note that with this assumption the model guarantees that mass is conserved when we form mixtures and compounds and when we break them up. To account for the law of constant proportions displayed by compounds, we must add one more assumption to our model. We must assume that when a compound is formed, each atom of one element attaches to a fixed number of atoms of the other elements in a pattern characteristic of the compound. A mixture apparently does not have such a characteristic pattern, and so the composition is not fixed.

According to our model, then, it is this attachment of atoms of different kinds to each other that is responsible for compounds having characteristic properties different from those of the elements of which they are made.

To get a better picture of what we believe is happening to individual atoms when a compound is formed, you can do the experiment described in the next section.

---

4 Suppose that *M* atoms of mercury combine with *N* atoms of oxygen to form mercury oxide.

a) What total number of atoms would you expect there to be in the sample of mercury oxide produced?

b) If the mercury oxide produced is then decomposed by heating to form gaseous mercury and oxygen, how many atoms of mercury and how many atoms of oxygen would you expect to find?

c) Would your answers to (b) be different if you condensed the samples of gas to liquid mercury and liquid oxygen?

8.4 "EXPERIMENT" FASTENERS AND RINGS; CONSTANT COMPOSITION

Since we cannot see atoms directly, the atomic model of matter may appear unreal to you. This experiment is designed to help you illustrate some aspects of it.

You will be using two "elements." The atoms of element Fs are paper fasteners; the atoms of element R are rubber rings. Note that the fasteners are all alike, just as we assume the atoms of one element are all alike. Similarly, all the rings are alike but differ from the fasteners because they are meant to represent atoms of another element (Fig. 8.2). We can fit rings on fasteners to "synthesize" a "compound" of two elements. When we do this, the numbers of rings and fasteners and their individual masses do not change. This agrees with the model.

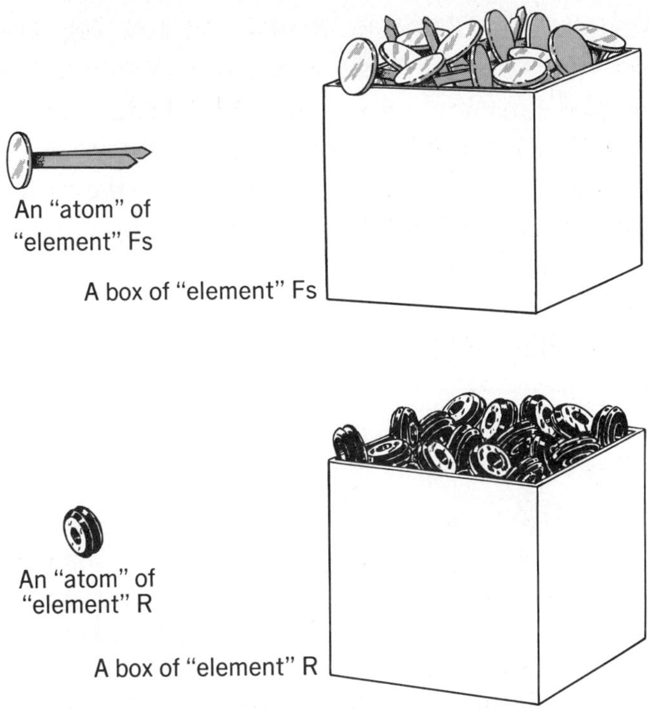

FIG. 8.2
Paper fasteners can be used to represent the atoms of "element" Fs, and rubber rings to represent the atoms of "element" R.

When you make a compound using these objects, you can think of the process as being similar to the formation of copper oxide from hot copper and the oxygen in the air. In that experiment you worked with large numbers of atoms that you could not see individually. In this experiment you will illustrate the process by using many visible "atoms" to form a sample of a compound.

Find the mass of all the Fs that you have been given. Make as much of the compound FsR as your supply of atoms allows by putting one ring on each fastener (Fig. 8.3). When we write FsR, we mean a compound made up of one atom of Fs for every atom of R. What is the mass of the product you have synthesized? If you have an excess of Fs, find its mass. How much Fs reacted with R?

What is the ratio of the mass of R to the mass of product in your sample of the compound? Compare your results with those of

FIG. 8.3
Making the "compound" FsR.

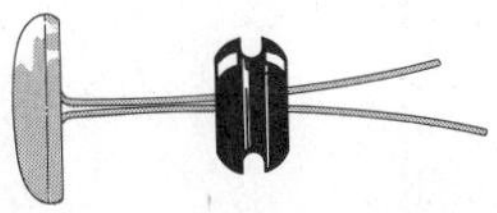

your classmates. Does this ratio depend on how big a sample you make? Does the model as illustrated by fasteners and rings agree with the law of constant proportions? Would the ratio have been the same if you had used heavier rings?

If you "decomposed" the compound you have made into pure Fs and R by taking the rings off the fasteners, you would get back all the atoms you started with. If you then measured the masses of the elements Fs and R that you got, would the model as illustrated by rings and fasteners agree with the law of conservation of mass?

---

5 **Does the experiment with rubber rings and paper fasteners give you any evidence for the shape of an atom?**

6 **In the synthesis-of-water experiment described in Sec. 6.3, different amounts of hydrogen were mixed with the same amount of oxygen, and then the mixture was ignited by a spark. In the first case there was some oxygen left over; in the second case nearly all the oxygen and hydrogen reacted; and in the third case hydrogen was left over. Describe the results of these experiments in terms of the atomic model of matter.**

7 **Suppose that element A can form a compound with element B but not with element C. Element B can form a compound with C. How would you choose fasteners, rings, and washers to represent these elements?**

8 **Describe the synthesis of zinc chloride (Expt. 6.4) in terms of fasteners, rings, and washers. (Hint: Hydrochloric acid contains hydrogen and chlorine.)**

8.5 "EXPERIMENT" SOME OTHER COMPOUNDS OF FS AND R

The compound FsR is only one of many that we can form from the elements Fs and R. An entirely different compound can be made from these elements by using two atoms of R for every atom of Fs. This new compound has the formula FsR_2 (Fig. 8.4). The $_2$ in the

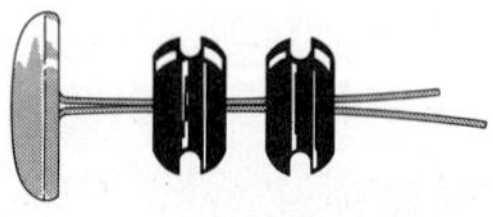

FIG. 8.4
Making the "compound" FsR_2.

formula means that there are two atoms of R for every atom of Fs in the compound.

Make a quantity of the compound FsR_2, decompose it, and find the total mass of R that combined with the total mass of Fs. What mass of R would have combined with 100 g of Fs? Using the data from the preceding experiment, what mass of R would have combined with 100 g of Fs in the compound FsR? How do the quantities of R that combined with 100 g of Fs in the two compounds FsR and FsR_2 compare? Does this ratio depend on the mass of the compounds you made?

Make a quantity of the compound FsR_3. Decompose it and determine the mass of R that combined with 100 g of Fs. Compare this mass of R in FsR_3 with the masses of R in FsR and FsR_2 that combine with 100 g of Fs.

9† The ratio of the mass of lead to the mass of oxygen in an oxide of lead is 13. How many grams of lead will combine with 100 g of oxygen?

10 You can make several "compounds" by gluing pennies (Pe) and nickels (Ni) together. The ratio of the mass of Ni to Pe in the compound NiPe is 1.6. What will be the ratio of the mass of Ni to Pe in the compound (a) Ni_2Pe, (b) Ni_3Pe, (c) $NiPe_2$?

A PREDICTION FROM THE ATOMIC MODEL OF MATTER 8.6

In the last two sections you synthesized compounds from fasteners and rings. The ratio of the masses of the elements that combined to form a given compound was constant, regardless of the original masses of the elements used. Of course, instead of fasteners and rings, you could have used magnets and nails, or nickels and pennies glued together, or almost any pairs of things to represent two different kinds of atoms. The detailed shapes and sizes and colors of our atoms were unimportant. The important point is that we imagine different kinds of particles that we can combine together.

The experiment you have just done allows us to make a prediction from the atomic model. Consider the results of the last fastener-and-ring experiment. You were asked to compare the masses of R that combine with 100 g of Fs in two compounds of these elements: FsR and FsR_2. You found their ratio to be close to 2.

You could have predicted this because all the rings have the same mass, all the fasteners have the same mass, and all the compound FsR contains only one ring for each fastener, whereas the other compound, FsR_2, contains two rings for every fastener.

What does the model predict about real compounds? Let us assume that we have two different compounds made only of the elements copper and chlorine. Suppose that one compound contains two atoms of chlorine for every atom of copper and the other compound contains only one atom of chlorine for every atom of copper. Then we expect the mass of chlorine combined with 100 g of copper in the first compound to be twice the mass of chlorine that is combined with 100 g of copper in the second compound.

Consider another possibility: One compound contains three atoms of chlorine for every atom of copper, and the other compound contains equal numbers of copper and chlorine atoms as before. (This would resemble the situation of FsR_3 and FsR.) Then you would expect the mass of chlorine in the first compound to be three times as great as in the second compound for a given mass of copper.

---

11 Two samples of two compounds containing only nuts and bolts were decomposed. A mass of 100 g of each sample yielded the following data.

Compound sample	Mass of bolts (g)	Mass of nuts (g)
1	80	20
2	67	33

a) What mass of nuts would combine with 100 g of bolts in sample 1?

b) What mass of nuts would combine with 100 g of bolts in sample 2?

c) What is the ratio of the mass of nuts in sample 2 to the mass of nuts in sample 1 that would combine with 100 g of bolts?

8.7 EXPERIMENT TWO COMPOUNDS OF COPPER

Copper and chlorine do in fact form two different compounds. Both compounds are solids at room temperature. One is brown, has a melting point of 498°C, and has no boiling point because the liquid decomposes before it is hot enough to boil. The other compound of copper and chlorine is light green and has a melting point of 422°C and a boiling point of 1366°C. We shall investigate these two compounds to make a comparison between the mass of chlorine that combined with 100 g of copper in one and the mass of chlorine that combined with 100 g of copper in the other.

You will be placing a piece of aluminum in two solutions, each containing a measured mass of one of these compounds. In each case the aluminum will replace the copper in the solution. The copper will precipitate, and you can separate it, dry it, and weigh it. Then you can compare the masses of chlorine that combine with 100 g of copper in the two compounds.

Find the mass of a watch glass as accurately as you can. Onto the watch glass put between 2 g and 4 g of the brown powder. Again mass very accurately the watch glass with the brown powder on it. Dissolve the powder in a beaker containing 25 cm^3 of water, and heat the solution to around 60°C. Now add a piece of aluminum to the solution. Stirring the solution from time to time with a glass rod will hasten the reaction.

When all the copper has been removed from the compound, the solution will be colorless. The solution can then be boiled a few minutes to loosen any copper deposited on the aluminum. Agitating the aluminum in the solution will help break off any bits of copper sticking to it. It may be necessary to scrape the aluminum with a scoopula to remove all the copper. Now you can remove what is left of the piece of aluminum.

When the copper has sunk to the bottom, you can pour off most of the liquid without losing any copper. You can wash out any of the remaining solution by adding 50 cm^3 of water and stirring thoroughly. To be sure none of the aluminum chloride solution remains, pour off the wash water. Again add 50 cm^3 of fresh water and stir, finally pouring off the water a second time. While washing, be sure to break up the copper chunks as much as possible with a glass rod. This will help remove any solution trapped in the spongy copper. If you wash it again with about 25 cm^3 of alcohol or burner fuel, the copper will dry rapidly. (Alcohol evaporates quickly.)

Now you can carefully scrape the copper onto the watch glass you massed earlier. To dry the copper you can either let it stand overnight or place the watch glass over boiling water for 20 to

30 min. When you think the copper is dry, mass the watch glass with the copper and then check the mass again after 10 min of additional heating to be sure the copper is dry. How can you tell if it is dry?

What mass of copper did you get? To see that the solid you have separated from the compound is copper, you can squeeze it into a lump and hammer it. Does it resemble copper?

What was the mass of chlorine in the sample of brown powder you used? What mass of chlorine is combined with 100 g of copper in this compound?

Repeat the experiment you have just done, but this time use 2 g to 4 g of the light-green compound of copper. What is the mass of chlorine that combines with 100 g of copper in this compound?

How does the mass of chlorine combined with 100 g of copper in the brown compound compare with the mass of chlorine combined with 100 g of copper in the light-green compound? What do the results of this experiment tell you about the number of chlorine atoms that combine with one copper atom in the two compounds?

---

12 In the experiment you have just done, how would the value for the mass of chlorine combining with 100 g of copper in the brown powder have been altered if this powder had not been thoroughly dried before massing?

13 In the case of the brown powder, suppose that some copper had been left in the beaker after the wet copper was transferred to the watch glass. How would this error have affected the results of the experiment?

14 A student repeated the experiment with the brown copper chloride several times, and each time he used the same strip of aluminum. In his last determination he found that there was no aluminum left to recover. How would this affect his value for the mass of chlorine combined with 100 g of copper in the last determination?

8.8 THE LAW OF MULTIPLE PROPORTIONS

There are other combinations of fasteners and rings that we can put together to illustrate possible compounds. Two of these, along with two of the combinations you have already made, are shown in Table 8.1. In each case we have calculated the mass of rings that would combine with 100 g of fasteners. The ratio of the masses is expressed as the ratio of small whole numbers.

With fasteners and rings we have illustrated some possible combinations of elements; now let us look at some actual compounds. Nitrogen and oxygen combine to form several oxides with distinctly different properties. The composition of these oxides has been determined. From the results shown in Table 8.2, you can see that the mass ratios are the same as in Table 8.1.

Table 8.1

"Compound formula"	Mass M of R combined with 100 g of Fs (g)	Mass ratios
FsR	$M_1 = 29.3$	—
FsR_2	$M_2 = 58.6$	$\frac{M_2}{M_1} = \frac{2}{1}$
Fs_2R	$M_3 = 14.7$	$\frac{M_3}{M_1} = \frac{1}{2}$
Fs_2R_3	$M_4 = 44.0$	$\frac{M_4}{M_1} = \frac{3}{2}$

These ratios are good evidence that elements do combine as though they were made up of individual particles.

Oxygen is known to combine in different mass ratios with many elements besides nitrogen. However, it is by no means the only element that behaves this way. Table 8.3 shows several other

Table 8.2

Compound: Name	Compound: Formula	Mass M of oxygen combined with 100 g of nitrogen (g)	Mass ratios
Nitrogen oxide	NO	$M_1 = 114.3$	—
Nitrogen dioxide	NO_2	$M_2 = 228.6$	$\frac{M_2}{M_1} = \frac{2}{1}$
Dinitrogen oxide	N_2O	$M_3 = 57.2$	$\frac{M_3}{M_1} = \frac{1}{2}$
Dinitrogen trioxide	N_2O_3	$M_4 = 171.5$	$\frac{M_4}{M_1} = \frac{3}{2}$

pairs of elements and their compounds. Note that the mass ratios in the last column are all expressed by small whole numbers. The results of the experiment you did with the chlorides of copper and of many other experiments on the composition of compounds also show mass ratios that are small whole numbers. Such simple ratios occur with compounds of two elements in which different masses of one element combine with a given mass of another element to form two or more compounds. These different masses have a ratio that can be expressed by two small whole numbers. This generalization is known as the law of multiple proportions.

Table 8.3

Compound	Formula	Composition	Mass ratios
Iron dichloride	$FeCl_2$	100 g iron, 127 g chlorine	$\frac{127}{191} = \frac{2}{3}$
Iron trichloride	$FeCl_3$	100 g iron, 191 g chlorine	
Methane	CH_4	100 g carbon, 33.3 g hydrogen	$\frac{33.3}{25.0} = \frac{4}{3}$
Ethane	C_2H_6	100 g carbon, 25.0 g hydrogen	
Phosphorus trichloride	PCl_3	100 g phosphorus, 344 g chlorine	$\frac{344}{573} = \frac{3}{5}$
Phosphorus pentachloride	PCl_5	100 g phosphorus, 573 g chlorine	$\frac{573}{229} = \frac{5}{2}$
Diphosphorus tetrachloride	P_2Cl_4	100 g phosphorus, 229 g chlorine	
Tin dichloride	$SnCl_2$	100 g tin, 59.9 g chlorine	$\frac{59.9}{120} = \frac{1}{2}$
Tin tetrachloride	$SnCl_4$	100 g tin, 120 g chlorine	

In predicting the law of multiple proportions from the atomic model of matter, we have been following in the footsteps of John Dalton (1766–1844). He did not use fasteners and rings in making his predictions, but he definitely thought in terms of two kinds of particles. He then found experimental evidence for the law of multiple proportions by analyzing compounds containing only oxygen and nitrogen (Table 8.2) and compounds containing only hydrogen and carbon.

Let us emphasize once more that we did not have to use fasteners and rings to predict the law of multiple proportions. We could just as well have used pennies and nickels stuck together with glue, because all pennies are alike, all nickels are alike, and nickels are different from pennies.

You have now studied two laws concerning the formation of compounds: the law of constant proportions (Sec. 6.5) and the law of multiple proportions. The first law applies to a single compound of any two given elements. To describe the formation of more than one compound by the same two elements, we need the second law. Thus the law of multiple proportions is an extension of the law of constant proportions.

When two elements can form more than one compound, it is often difficult to synthesize only one compound without forming some of the others at the same time. The relative amounts of the various compounds depend on the relative quantities of the elements with which we start, the temperature, and other conditions. In such cases experiments will yield different ratios for the elements forming the compound. Part of the difficulty of establishing the law of constant proportions had its origin in the formation of more than one compound by the same two elements. This is one of the reasons why the combining of copper and sulfur seemed to violate the law of constant proportions. (See Sec. 6.5.) Another complication that occurs when sulfur is in excess is that the compounds formed dissolve in the excess sulfur to form "frozen" solutions, thus making it very difficult to separate the compounds.

Note that the law of multiple proportions does not tell us what specific ratios to expect from any given pair of elements that form more than one compound. It states only: If two elements A and B form two or more compounds, then the ratio of the masses of A combining with a given mass of B will be given by the ratio of two small whole numbers. Some pairs of elements form several compounds whereas others form only one or even none (helium, for example, is not known to combine with any other element). This means that there must be some important differences between the atoms of the various elements to account for their different behavior in forming compounds.

15† What is the ratio of the mass of chlorine that combines with 100 g of phosphorus in phosphorus trichloride to the mass of chlorine that combines with 100 g of phosphorus in diphosphorus tetrachloride (Table 8.3)?

16† From the data given below for compounds of lead and oxygen, calculate the ratios in the last column. Do these ratios agree with the law of multiple proportions?

Compound	Formula	Mass M of oxygen combined with 100 g of lead (g)	Mass ratio
Lead (I) oxide	Pb_2O	$M_1 = 3.86$	—
Lead (II) oxide	PbO	$M_2 = 7.72$	$\frac{M_2}{M_1}$
Lead (III) oxide	Pb_2O_3	$M_3 = 11.58$	$\frac{M_3}{M_1}$
Lead (IV) oxide	PbO_2	$M_4 = 15.44$	$\frac{M_4}{M_1}$

17 Two compounds containing only carbon and oxygen are decomposed. A mass of 100 g of compound I contains 43 g of carbon, and 100 g of compound II contains 27 g of carbon.

a) What is the ratio of the mass of carbon to the mass of oxygen for each compound?

b) If compound II has the formula CO_2, what is a possible formula for compound I?

8.9 MOLECULES

So far, we have described the formation of compounds in terms of the atomic model in a rather vague way: In every compound made of two elements *A* and *B*, the atoms arrange themselves in such a way that for each atom of *A* there are one or more atoms of *B*. We have not yet attempted to describe exactly how large numbers of these are arranged in a sample of a compound. Figure 8.5 shows several possibilities for a compound containing equal numbers of atoms of two elements. In Fig. 8.5(*a*) the atoms of the two elements are clearly paired off with space between each pair. In Fig. 8.5(*b*) the atoms form clusters of four. In Fig. 8.5(*c*) and (*d*) we can no longer distinguish groups or clusters containing a fixed number of atoms, but each sample contains an equal number of atoms of each kind. Which of these pictures fits the arrangement of atoms in real

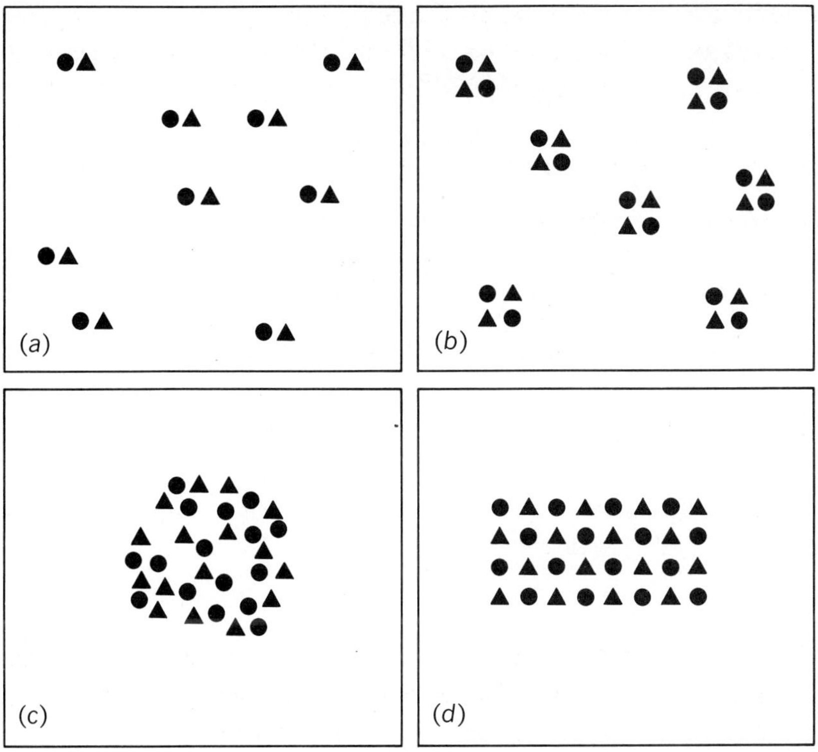

FIG. 8.5
Possible arrangements of the atoms of two elements in a compound that contains equal numbers of each kind of atom.

compounds? To answer this question, we must look into what we already know about the bulk behavior of matter.

Anybody who has ever used a bicycle pump knows that it is quite easy to compress air into a volume much smaller than the volume it originally occupied. The same is true for all gases. But try to reduce the volume of a small crystal of sodium chloride by only 1 percent by squeezing it in a cylinder with a piston; you will not be able to do it without special equipment that can exert very large pressures. A cube of copper will behave similarly. While gases are quite compressible, solids are almost incompressible. We can account for this by assuming that atoms behave like little hard objects whose size remains fixed. In a solid the atoms are close together; they are all "touching" one another. This means that the distance between the centers of the atoms equals the size of the atoms themselves (Fig. 8.6). Such a picture suggests that solids are hard to compress. On the other hand, in a gas the atoms do not touch one

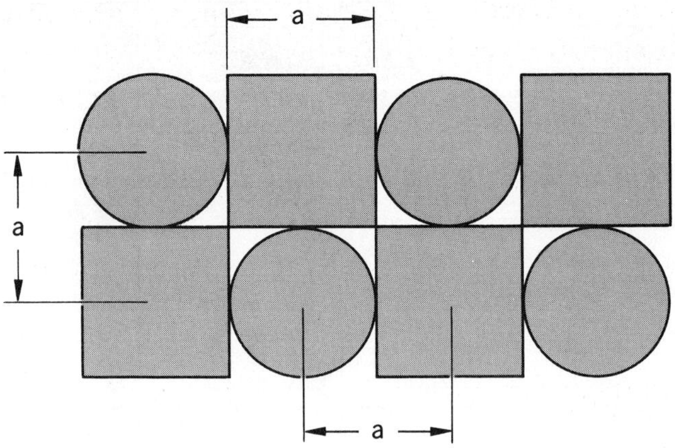

FIG. 8.6
In a solid the distance between the centers of adjacent atoms equals the size of the atoms themselves.

another. They are far apart compared with their size and can easily be pushed closer together to occupy a smaller volume (Fig. 8.7). Thus, as a solid or liquid the compound *AB* can be represented by Fig. 8.5(*c*) or (*d*), whereas the same compound is represented as a gas by Fig. 8.5(*a*) or (*b*). Groups or clusters containing a fixed number of atoms, like those in Fig. 8.5(*a*) and (*b*), are called "molecules." Each molecule of a given compound contains the same number of atoms.

Compounds are not the only substances made of molecules. The atoms of a single element can also form clusters or molecules in which there is only one kind of atom in each cluster. In this course we shall not go into the analysis of the many experiments and the long chain of arguments that are necessary to decide the numbers of various kinds of atoms in the molecules of different gases. We shall simply state the results for some of the substances we have used as illustrations: In gaseous hydrogen the atoms cluster together in pairs; thus the correct molecular formula of hydrogen gas is H_2. The atoms of gaseous iodine also cluster together in pairs. A molecule of water vapor (H_2O) consists of two atoms of hydrogen and one of oxygen, and a molecule of carbon dioxide (CO_2) consists of a cluster of two oxygen atoms and one carbon atom. The atoms of helium gas (He) do not form clusters.

When the atoms of a solid are arranged as in Fig. 8.6, it is meaningless to talk about clusters or molecules unless we wish to consider an entire crystal as one huge molecule. The only information a formula gives us in such cases is the fixed ratio of the numbers of the various atoms forming the compound. Thus the formula NaCl, when applied to solid sodium chloride, means only that the crystal contains equal numbers of sodium atoms and chlorine atoms. On the other hand, a molecule of sodium chloride gas (the liquid boils at 1400°C!) exists and is correctly described by the for-

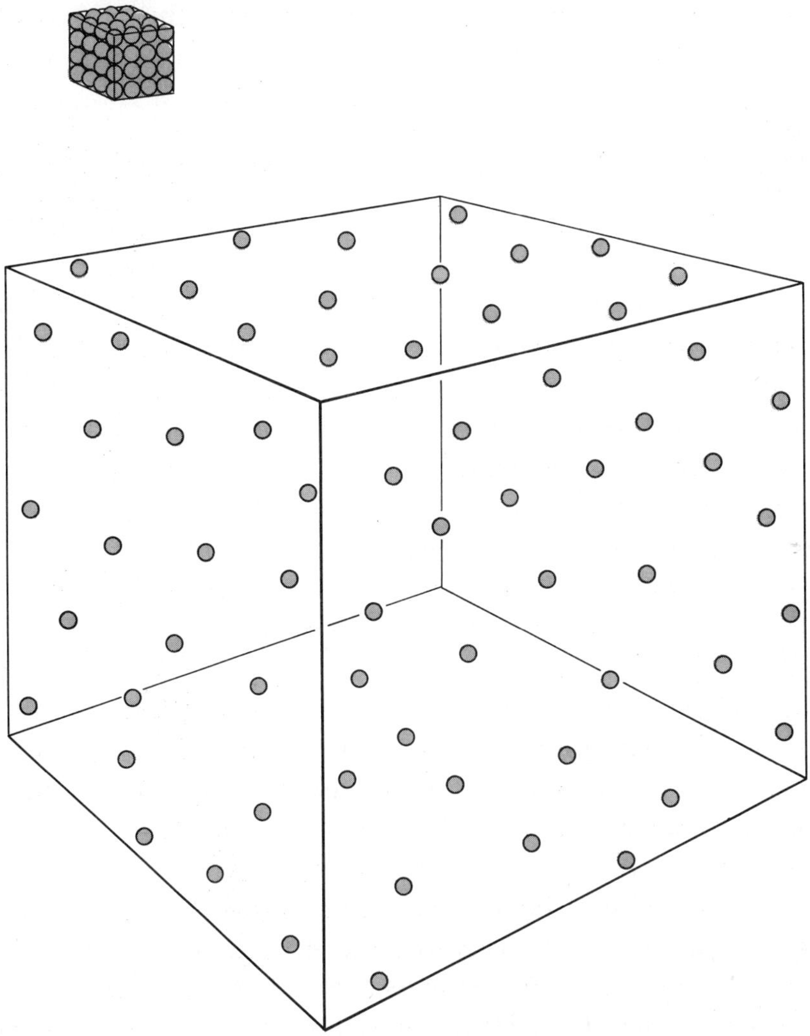

FIG. 8.7
The large box represents a volume of helium gas at atmospheric pressure and room temperature. The small spheres in the box represent helium atoms. The small box shows the volume occupied by the helium atoms in the large box when the helium is liquefied.

mula NaCl. Each molecule of the gas is made of one atom of sodium and one atom of chlorine.

A closer look at solids shows that they are all practically incompressible when compared with gases. However, some solids are more compressible than others. This suggests that perhaps, in some of the more compressible solids, not all the atoms touch one another but that the atoms do form clusters or molecules and the empty space between clusters accounts for the larger compressibility. For example, iodine crystals are quite a bit more compressible than those of copper. And there is indeed evidence that the iodine crystal is made up of pairs of atoms. The atoms in a pair have their centers closer together than atoms in separate pairs. Thus it is meaningful to speak of a molecule of solid iodine. Liquids certainly are incompressible when compared with gases, but they are more compressible than most solids, again indicating the possible existence of distinct molecules.

---

18 Why is it more difficult to speak in terms of molecules in a solid than in a gas?

19 What experimental evidence indicates that some solids do consist of a collection of molecules?

8.10 RADIOACTIVE ELEMENTS AND THE ATOMIC MODEL

The basic idea in the atomic model of matter, as we have developed it, is that in forming mixtures and compounds the number of different kinds of atoms and their individual masses do not change. In this chapter we have seen the usefulness of this idea. But what about radioactive decay?

We can extend our model to include radioactivity by assuming that in radioactive decay an atom of the decaying element splits into two atoms of the new elements being formed (lead and helium, in the case of polonium). This assumption does not really contradict our basic idea that the number of atoms remains fixed. This idea still applies to such processes as melting, evaporating, mixing, forming compounds, and breaking them up. The one exception is the special process of radioactive decay. The assumption that in radioactive decay an atom splits into two atoms is forced upon us by our observation of the process.

This state of affairs may not seem entirely satisfactory, but it is a common situation in the development of science. No sooner do we arrive at a model that seems to put our knowledge of nature in order than we begin to discover its limitations. We find that further experiments do not fulfill the predictions based on the new theory, and so we must modify and expand it.

For Home, Desk, and Lab

20 Think again of the sealed tin can referred to in Sec. 8.1. You are not allowed to pierce the can or break it open. What would you predict about its behavior if (a) you lowered the temperature sufficiently or (b) you raised the temperature sufficiently?

21 While doing Expt. 8.2, A Black Box, why were you not permitted to open the box and look inside?

22 Which of the following are ***parts*** of our atomic model of matter? Which may be used to ***illustrate*** the model?

a) Matter is made up of very tiny particles, much too small to be seen.

b) A paint sprayer will deposit more particles the longer it is operated.

c) Atoms of the same element are all alike.

d) Two rubber rings combine with one fastener to form FsR_2.

e) Atoms of two different elements may combine to form compounds.

f) Atoms of two different elements may combine in different ratios to form different compounds.

g) Marbles can be stacked in a box to represent a solid.

23 A mass of 16 g of oxygen combines with 63.5 g of copper to form CuO.

a) What is the ratio of the mass of copper to the mass of oxygen?

b) What is the ratio of the mass of copper to the mass of oxygen in Cu_2O?

24 In your own words state the law of multiple proportions.

25 Carbon can combine with chlorine to form three different compounds, CCl_4, C_2Cl_4, and C_2Cl_6. If the ratio of the mass of carbon to chlorine in CCl_4 is 0.286, what is the ratio of the mass of carbon to that of chlorine in the other two compounds?

26 Suppose that 6×10^6 atoms of hydrogen combine with 2×10^6 atoms of nitrogen and form 2×10^6 molecules of ammonia. How many atoms of hydrogen and how many atoms of nitrogen would there be in each molecule of ammonia?

27 A student suggested the following formulas for the two chlorides of copper.

Brown Chloride	Green Chloride
$CuCl_2$	$CuCl$
$CuCl$	Cu_2Cl
Cu_3Cl_2	$CuCl$
Cu_3Cl_4	Cu_3Cl_2

a) Which pair or pairs are possible formulas for these compounds?
b) Can you decide which of the pairs gives the correct formulas for the chlorides of copper?

28 How does the atomic model of matter enable us to account for (a) the law of constant proportions, (b) the law of multiple proportions, and (c) the law of conservation of mass?

Molecular Motion

9

MOLECULAR MOTION AND DIFFUSION 9.1

When we put a piece of wood on the table, it stays there. But as soon as we take the cap off a bottle of ammonia solution, some of the gas comes out of the solution, escapes into the air, and spreads in all directions. On a cold winter day some of the hot water in a bathtub evaporates and spreads throughout the room, condensing on the cold windowpanes and walls. Water vapor, a gas, moves from the tub in all directions. If a gas spreads all by itself, then we have to conclude that its molecules must be in motion.

To demonstrate that the molecules of a gas move around freely, we find it convenient to use a gas that is visible and that can be liquefied easily. The element bromine has both these properties: It has a deep red-brown color and will solidify when cooled by a mixture of Dry Ice and alcohol.

The photograph on the far left in Fig. 9.1 shows a sealed glass tube containing a small quantity of bromine and, of course, air. The tube was cooled by a mixture of Dry Ice and alcohol. All the bromine is seen as a solid at the bottom of the tube.

The other photographs, reading from left to right, show the tube at different times while it was warming. You can clearly see

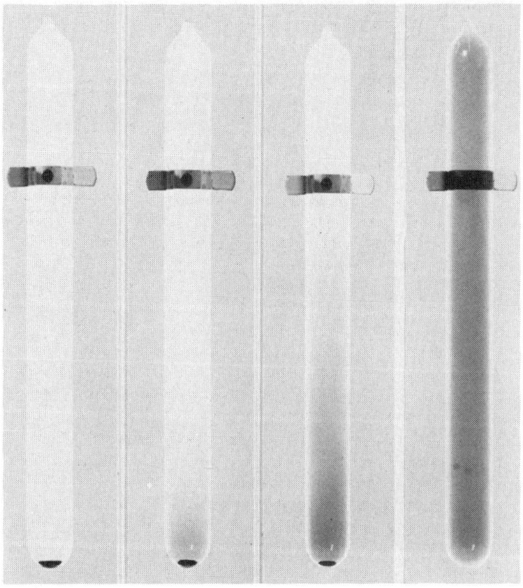

FIG. 9.1
A series of photographs, taken at different times, of a sealed glass tube containing bromine. In the photograph at the far left, taken immediately after the tube was removed from a mixture of Dry Ice and alcohol, the bromine is all at the bottom of the tube as a solid. As the tube warms up, the bromine turns to liquid and evaporates. Brown bromine gas slowly spreads upward, gradually darkening the tube until it is evenly distributed.

that, as the bromine evaporates, it moves slowly upward until it is evenly distributed throughout the tube. This set of pictures suggests that the bromine molecules must be moving around very slowly.

To be sure that the experiment shown in Fig. 9.1 really tells us something about the speed of the bromine molecules, we have to repeat the experiment with bromine and no air (or at least almost no air) in the tube.

Figure 9.2 shows a set of pictures of a tube containing a little bromine and very little air. Most of the air was pumped out, and the tube was sealed while the bromine was frozen. Then the tube was allowed to warm up in the same way as the tube in Fig. 9.1. Note that, almost as soon as even the faintest color is observed, the color is the same throughout the tube. The color gets darker as the tube warms up and more bromine evaporates; but if you covered up the two ends of the tube with your hands, you could hardly tell which end contained the solid bromine. Contrast this with Fig. 9.1, where there is no doubt which way the bromine is moving. We can there-

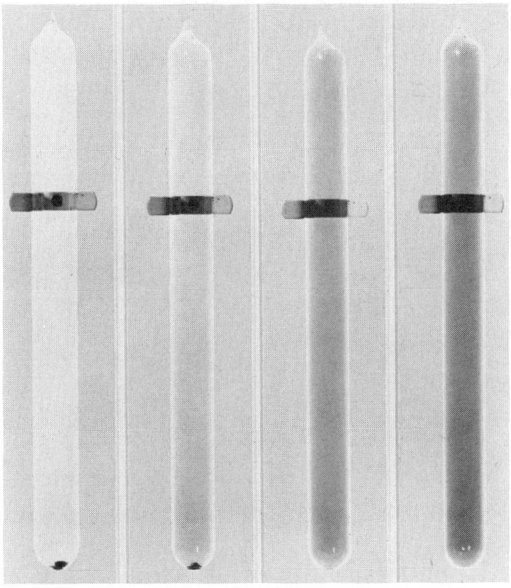

FIG. 9.2
The tube shown in this series of photographs is the same as that shown in Fig. 9.1 except that there is practically no air in it. It was photographed at the same intervals as the tube in Fig. 9.1. Note that as the bromine evaporates, it fills the tube evenly, coloring the whole tube a darker and darker shade. Unlike the case shown in Fig. 9.1, this time the gas does not move slowly up the tube.

fore conclude that when there is no air in the tube, the bromine molecules move so fast that they bounce back and forth and distribute themselves evenly inside the tube in a time too short for us to perceive.

We can now understand why the bromine molecules seemed to move so slowly in the tube containing air (Fig. 9.1). Think of a group of boys trying to spread out across a football field on which several hundred people are milling around. Even if a boy runs at top speed, he will move only a few steps before he bumps into someone. He may be pushed aside a little, or he may even be thrown backward. His next collision will have similar results. What would otherwise be a straight dash across the field becomes a zigzag path. It is no wonder that under such circumstances it will be a long time before even a few of the boys reach the other side of the field. In this picture the group of boys plays the role of the evaporating bromine; the milling crowd, that of the air in the tube.

To sum up, molecules of a gas apparently bounce around

at high speeds. Because of the many collisions, one gas spreads through another rather slowly. This process of one gas spreading through another is called *diffusion*.

1 What would be different about the photographs in Fig. 9.1 if (a) the bromine evaporated more rapidly and (b) there were more air in the tube?

2† What evidence leads you to believe that atoms are in motion and not at rest?

9.2 DENSITY AND PRESSURE OF A GAS

By using fasteners and rings to illustrate the atomic model, we were able to make a prediction about how elements combine to form compounds, and we confirmed this prediction with two chlorides of copper. When we used fasteners and rings, they were at rest. They were not in constant zigzag motion, as we know the molecules of a gas to be. If we wish to predict the properties of gases made up of rapidly moving molecules, we must extend the atomic model to include the motion of molecules. To illustrate the motion of the molecules in a gas and to help us predict some effects of this motion, we shall use small steel spheres. Of course, steel spheres do not stay forever in motion on their own. If we want to study the effects of bouncing steel balls over a period of time, we have to keep them in motion. We can do this with the apparatus shown in Figs. 9.3 and 9.4.

The spheres inside the cylinder (Fig. 9.3) are resting on a movable platform connected to an electric motor. A plastic disk is resting on top of the spheres. When the motor is turned on, the platform moves rapidly up and down over a short distance, keeping the spheres in motion.

Figure 9.5 shows what happened to the plastic disk when the motor was turned on. The shutter of the camera was open for 4 seconds. During this time the little spheres bounced around so many times that in the figure they have been smeared out and are practically invisible. Yet their presence is very much in evidence; note that the disk is now held at a position well up the cylinder by the moving spheres bouncing off the bottom of the disk.

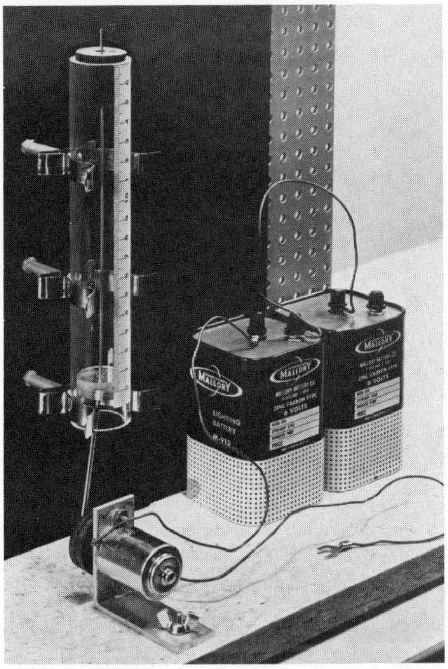

FIG. 9.3
A plastic cylinder, with a movable platform near the bottom, contains some small steel spheres and a movable disk lying on top of the spheres. The movable platform is vibrated up and down by a shaft connected to a small electric motor.

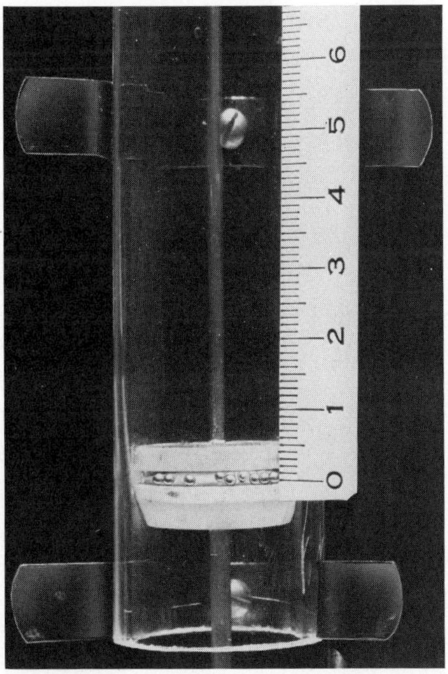

FIG. 9.4
A close-up view of the movable platform, spheres, and disk shown in Fig. 9.3.

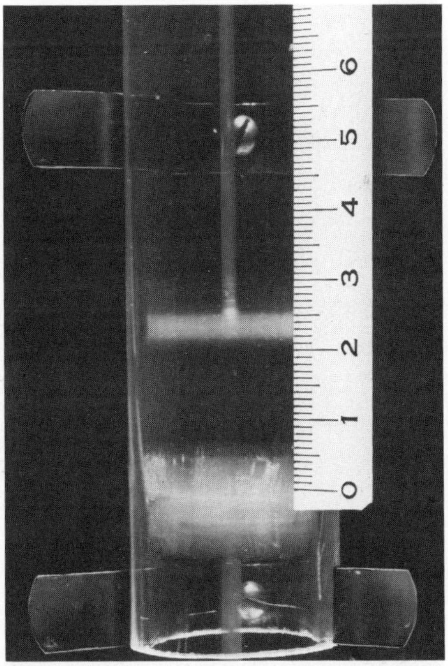

FIG. 9.5
The cylinder shown in Fig. 9.4, with the bottom platform being vibrated rapidly up and down over a short distance by an electric motor. The photograph is a time exposure of 4 seconds.

In Fig. 9.5 there are 10 spheres in the tube. How does the volume of the "sphere gas" change if we double the number of spheres in the tube? Figure 9.6 gives us the answer: The volume has increased; in fact, it has almost doubled. To get the spheres back to the volume they were occupying before we doubled their number, we have to add to the mass of the disk. With twice the mass pushing down on the spheres, we find that a sample of sphere gas containing 20 spheres occupies the same volume as a sample with 10 spheres held down by the original mass (Fig. 9.7). In other words, twice as many spheres bouncing around in the same volume push up on the disk twice as hard. This suggests that the contribution of each sphere to holding up the disk is independent of that of all the other spheres. If we double the number of spheres, while keeping the volume constant, they will be able to support twice the mass. Therefore doubling the number of spheres doubles the pressure they exert on the bottom of the disk. The mass of the spheres is not important. Doubling the number of light spheres or doubling the number of heavy spheres doubles the pressure.

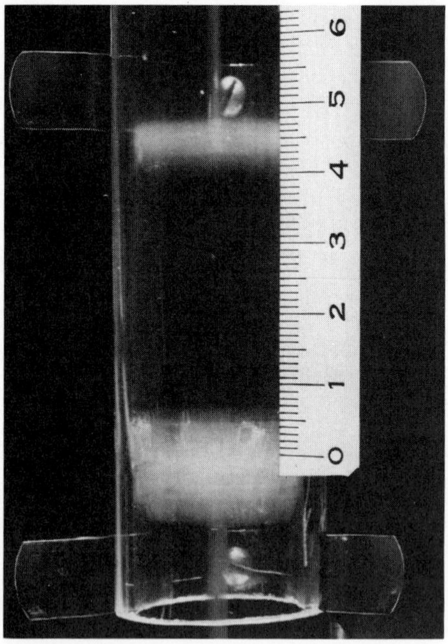

FIG. 9.6
With twice as many steel spheres in the cylinder as in Fig. 9.5, the volume of the "sphere gas" is close to twice as great.

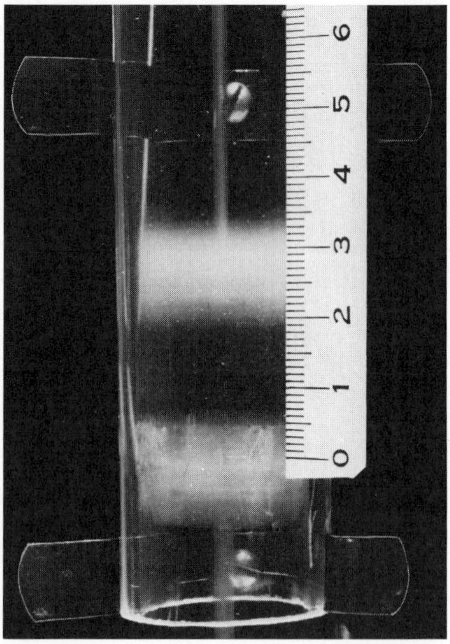

FIG. 9.7
Adding a second disk, whose mass equals that of the original disk, reduces the volume occupied by 20 spheres to that occupied by the 10 in Fig. 9.5.

3† If a gas is compressed until its pressure is doubled, what will happen to the density of the gas?

4 In the sphere-gas machine, what would you expect to happen if the top disk had less mass?

5 Two bricks are placed on each of three wood dowels resting on clay as shown in Fig. A. Which dowel will sink fastest into the clay? On which is the pressure greatest?

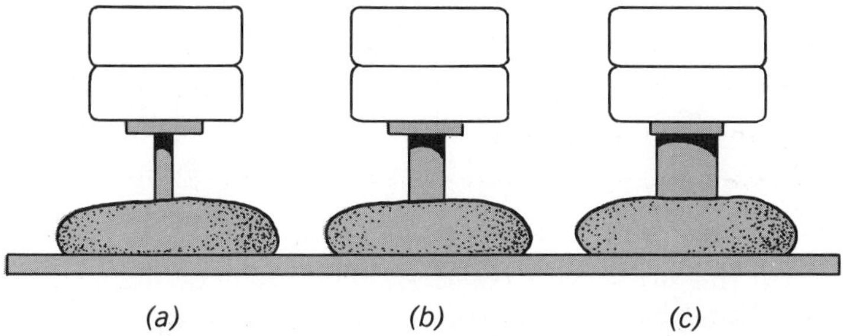

FIG. A For prob. 5.

6 Two bricks are placed on each of the pistons shown in Fig. B. In which cylinder will the pressure of the gas be greatest?

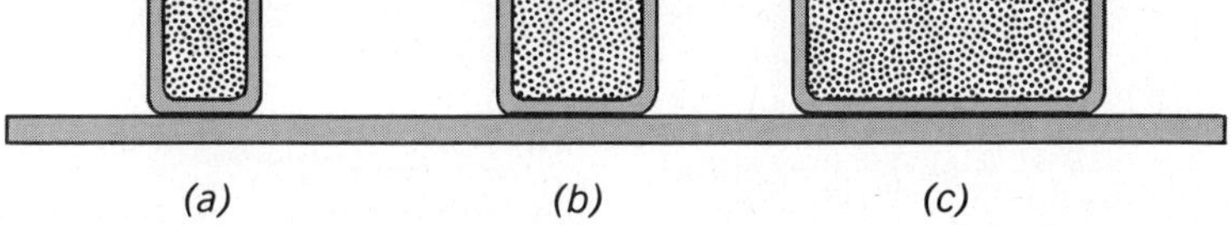

FIG. B For prob. 6.

A PREDICTION ABOUT THE RELATION BETWEEN VOLUME AND PRESSURE OF GASES 9.3

Suppose we double the number of molecules of a gas in a cylinder equipped with a piston like that shown in Fig. 9.8. The piston cor-

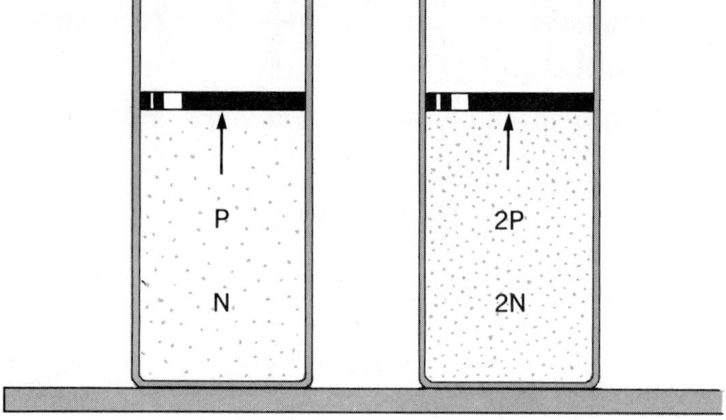

FIG. 9.8
(*a*) There are *N* molecules in the cylinder. (*b*) If we double the number of molecules in the cylinder and the piston is kept in the same position as in (*a*), the pressure *P* exerted by the gas will double.

responds to the disk in the sphere-gas machine. If we keep the volume constant (thereby doubling the density), the pressure exerted by the gas on the bottom of the piston will also double.

Forcing more gas into a given volume is not the only way to increase its density. We can also increase its density by keeping its mass constant and decreasing its volume. Consider a cylinder containing a gas (Fig. 9.9). In Fig. 9.9(*a*) there are 2*N* molecules (*N* molecules in the upper half and *N* molecules in the lower half.) If

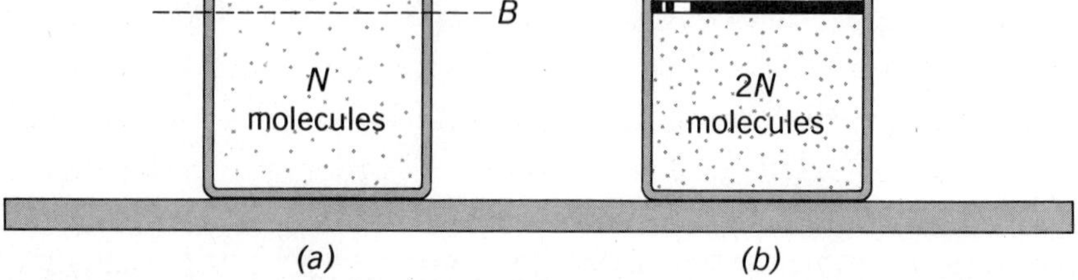

FIG. 9.9
Halving the volume of a gas by moving the piston down to *B* will double the density and, we predict, the pressure.

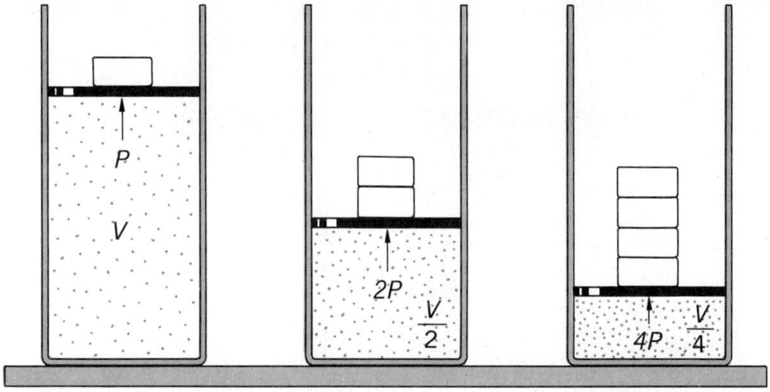

FIG. 9.10
The atomic model of a gas predicts that as the volume of a gas is decreased, its pressure rises in the manner shown by the three diagrams.

we push the piston down to *B*, we shall then have 2*N* molecules in the lower half (Fig. 9.9(*b*)). The density of the gas will have doubled; and on the basis of what we have seen in the preceding section, we can expect that the gas will exert a pressure that is just twice as great as when the piston was at the top of the cylinder.

If we push the piston down still farther so that it is halfway between *B* and the bottom of the cylinder, the density will double again. If our model still holds, the pressure that the molecules will exert on the piston will also double (Fig. 9.10). In other words, we can predict that when the volume of a gas is reduced to one-fourth its original volume, the pressure will increase to four times the original pressure.

We can generalize our predictions by saying that if we *decrease* the volume of a gas by a certain factor, the pressure of the gas will *increase* by the same factor. This relationship is illustrated in Fig. 9.11.

Note that, in developing our argument, we have made two basic assumptions: (a) The molecules of a gas are constantly in motion, and (b) the bouncing of one molecule off the wall of a container is not affected by the presence of other gas molecules. We made no assumptions as to how many atoms, or atoms of what elements, are in a molecule of the gas. Thus, if the predicted relation between pressure and volume of a gas is at all correct, it should be so for all gases as long as the two basic assumptions mentioned here are correct.

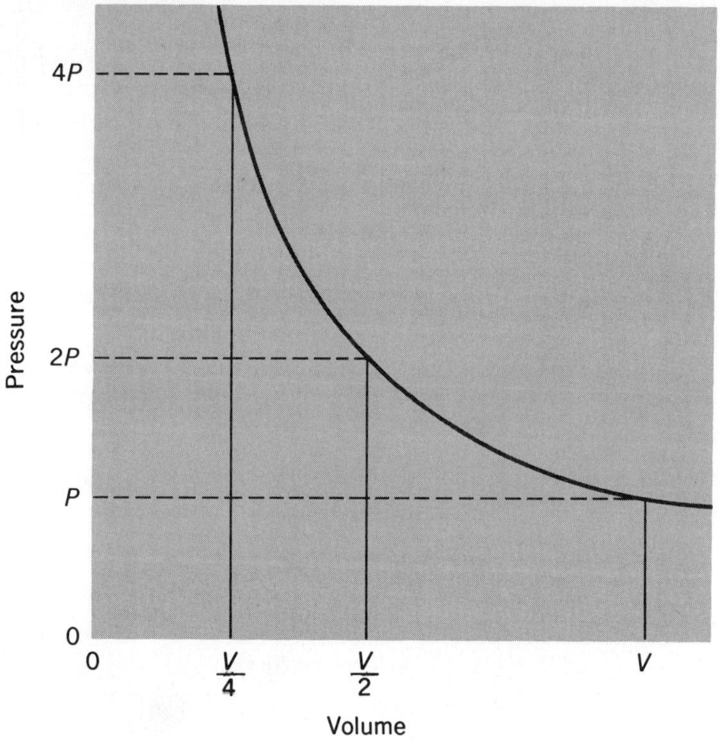

FIG. 9.11
A graph of the pressure of a gas as a function of its volume, as predicted from the atomic model.

9.4 THE COMPRESSIBILITY OF GASES

The apparatus shown in Fig. 9.12 was used to check the prediction we have just made about the relation between the pressure and the volume of air. By adding bricks to the platform on the top of the syringe, the air in the syringe was compressed.

Before we added bricks to the platform, we adjusted the initial volume of gas in the syringe to about 30 cm^3. With this volume the piston is stable enough to hold up to four bricks. We did this by pushing the piston down while a length of string was held alongside the piston seal (Fig. 9.13). The string allows air to escape as the piston is pushed down. When we reached the desired volume, we pulled out the string.

Our procedure is complicated by the effect of friction. When the piston is pushed down, friction prevents it from moving as far

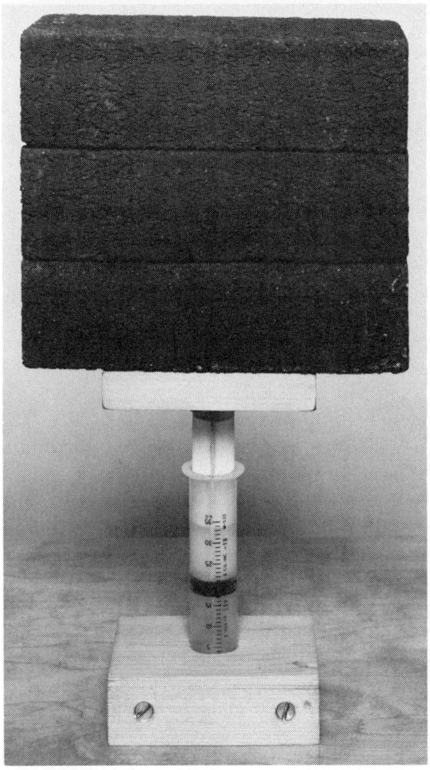

FIG. 9.12
A syringe can be used to find the relation between the pressure and the volume of a gas. The gas is compressed to smaller and smaller volumes by adding bricks to the platform on the top of the syringe.

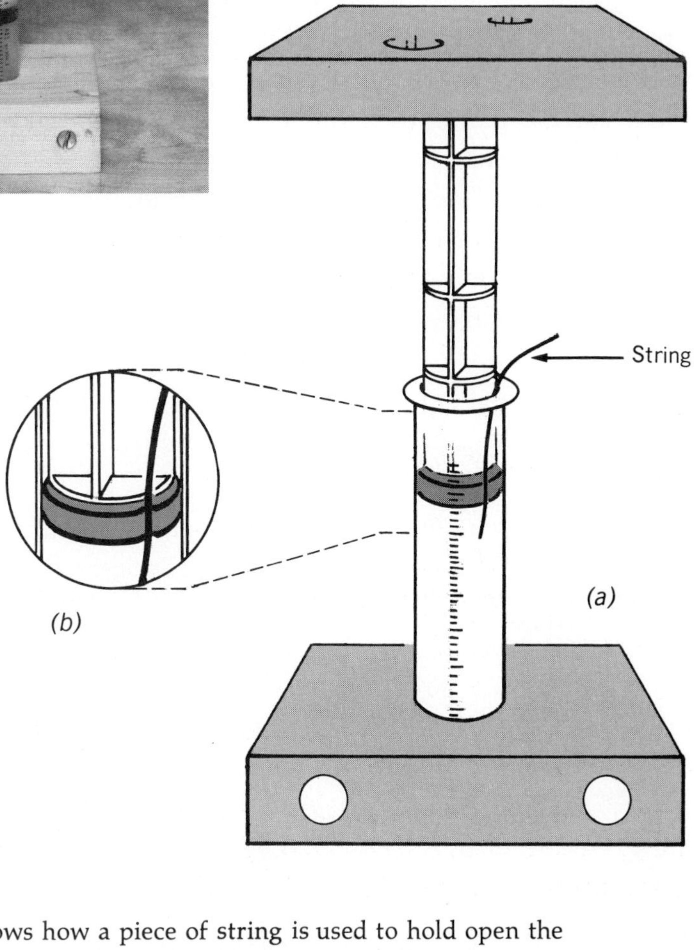

FIG. 9.13
The inset in (*b*) shows how a piece of string is used to hold open the seal of the syringe in (*a*) when adjusting the initial volume of air in the syringe.

down as it would if there were no friction. This results in a larger volume than we would expect in the absence of friction.

The reverse happens when we remove bricks. If we remove them, one by one, the sticking caused by friction will then prevent the piston from *rising* as far as we would expect. However, we can overcome the problem caused by friction if for each number of bricks on the platform we take two readings of each volume, one on the way down and one on the way up. Taking the average of each pair of readings very nearly cancels the error introduced by friction.

After adjusting the initial volume and before adding the first brick, we pulled the platform up a little and let it fall until it settled. We then recorded the volume for "zero bricks."

Next, we added one brick and, without giving it any downward push, let it come to rest and recorded the volume. We continued to add bricks in the same way until we reached a total of four bricks. At this point we had half our volume measurements—those for the piston as it moved down under the pressure of more and more bricks. To obtain volume measurements for the piston on the way up, we first pushed down slightly on the four bricks. After removing our hand, the piston rose, but nevertheless came to rest slightly below its resting point before we gave our push. This resting point gave us our second reading for the largest number of bricks.

We then removed the bricks one-by-one and recorded each volume in turn. The data we obtained are shown in Table 9.1.

Table 9.1

DOWN		UP		
No. of bricks	Volume (cm^3)	No. of bricks	Volume (cm^3)	Average volume (cm^3)
0	30.0	0	28.4	29.2
1	22.6	1	20.0	21.3
2	17.8	2	15.7	16.8
3	14.1	3	12.8	13.5
4	11.7	4	10.9	11.3

The graph of these data is shown in Fig. 9.14. It looks much like Fig. 9.11. Both show that the pressure increases when the volume decreases. But a careful examination shows that there is a significant difference: When the volume is halved from 25 to 12.5 cm^3, the pressure goes from about 0.5 bricks to 3.4 bricks, not up to 1.0 brick, as we should expect on the basis of Fig. 9.11. Similarly, when we increased the pressure from 1 to 2 bricks, the volume was reduced from 21.3 to 16.8 cm^3 and not to about 10.7 cm^3, as we might expect. Is the theory wrong?

Let us look again at the details of the experiment described in this section. We started with no bricks on the wooden platform and then went up to 1 brick, 2 bricks, and so on. But actually the bricks were not all that were pressing on the gas in the cylinder. The piston, the wooden platform on which the bricks were resting, and the air above the cylinder were all pressing on the gas. Thus, when we replace one brick by two bricks, we do not double the pressure.

To get the total pressure on the gas, we must add to the pressure exerted by the bricks the pressure caused by the platform, the piston, and the air. How much pressure, in terms of bricks, must

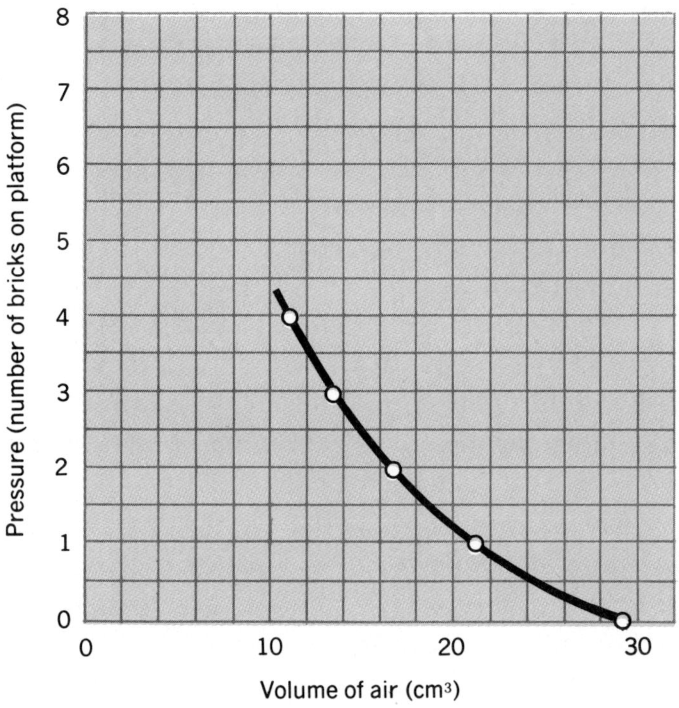

FIG. 9.14
The pressure of a gas (air) as a function of volume. The data are plotted from Table 9.1. Note that when the volume is halved, the pressure does not double.

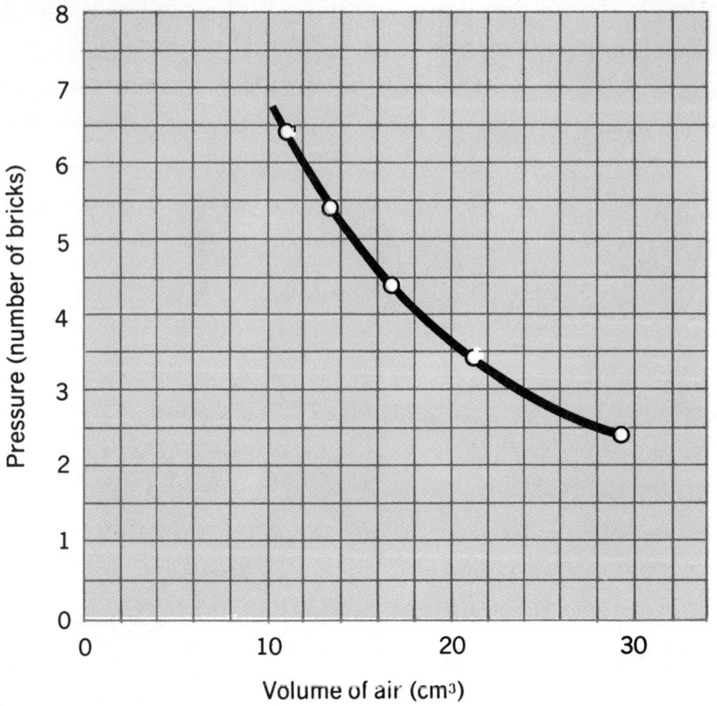

FIG. 9.15
The same graph as shown in Fig. 9.14 but with the curve moved vertically upward by 2.4 bricks on the pressure scale.

we add? Suppose that we must add one brick. This would raise every point on the curve one unit. Had we done this, we would have found it to be not enough.

By trial and error we found that raising each point on the curve of Fig. 9.14 by the equivalent of 2.4 bricks makes a curve that agrees closely with Fig. 9.11. Our result is shown in Fig. 9.15. In Fig. 9.15 doubling the pressure halves the volume. For example, when we go from 3 to 6 bricks, the volume decreases from 24.0 cm^3 to 12.1 cm^3, which is very close to halving the volume. Our prediction, made from the atomic model of a gas, is indeed borne out by our measurements of the pressure and the corresponding volume of air as it is compressed in a cylinder by different numbers of bricks.

The experiment has been repeated with other gases such as carbon dioxide and propane. If the initial volume is the same as in Fig. 9.14, and the same syringe and bricks are used, the graphs are the same for air, carbon dioxide, propane, and other gases. In other words, under the conditions of our experiment, different gases have the same compressibility. This is what the model predicted for the behavior of gases.

The relationship we have found between the pressure and the volume of a gas was discovered by Robert Boyle in 1662 and bears his name. It is called "Boyle's law." We predicted Boyle's law from the atomic model. Boyle discovered the law by doing experiments similar to the one we have described in this section. This was before there was an atomic model for gases from which Boyle's law could have been predicted.

7† From the graph in Fig. 9.11, what is the pressure when the volume is reduced (a) from V to ⅓ V, (b) from V to ¾ V?

8 In Fig. 9.15 when the total pressure is doubled by increasing from 2.5 bricks to 5 bricks, what is the ratio of the initial volume to the final volume?

9 A cylinder contains 1,000 cm^3 of hydrogen at a total pressure of 1.0 atmosphere. What will be the total pressure in atmospheres if the temperature does not change and the volume is reduced to (a) 100 cm^3? (b) If it is reduced to 10 cm^3?

10† A cylinder contains 100 cm^3 of air at a total pressure P (pressure due to piston plus atmospheric pressure). What will the total pressure become if the volume is reduced to (a) 50 cm^3, (b) 10 cm^3?

11 Figure 9.15 is a graph of pressure as a function of volume of air over a range of volume from about 30 cm^3 to about 10 cm^3. What would you predict the total pressure to be (in terms of bricks) if the volume of gas were (a) 10 cm^3, (b) 70 cm^3?

BEHAVIOR OF GASES AT HIGH PRESSURES 9.5

We have seen that gases have the same compressibility at pressures near atmospheric pressure. What happens if we compress gases to very high pressures?

Figure 9.16 is a graph of volume as a function of pressure for nitrogen and hydrogen at very high pressures and at 0°C. The lowest curve is the graph that could be expected for both gases if their behavior were correctly described by Boyle's law. You can see that both gases follow Boyle's law at pressures below approximately 150 atmospheres. It is impossible to distinguish between the two gases by their compressibility at low pressures.

At high pressures, however, the story is quite different. Neither gas is as compressible as one would expect from Boyle's law. At greater and greater pressures the compressibilities of the two

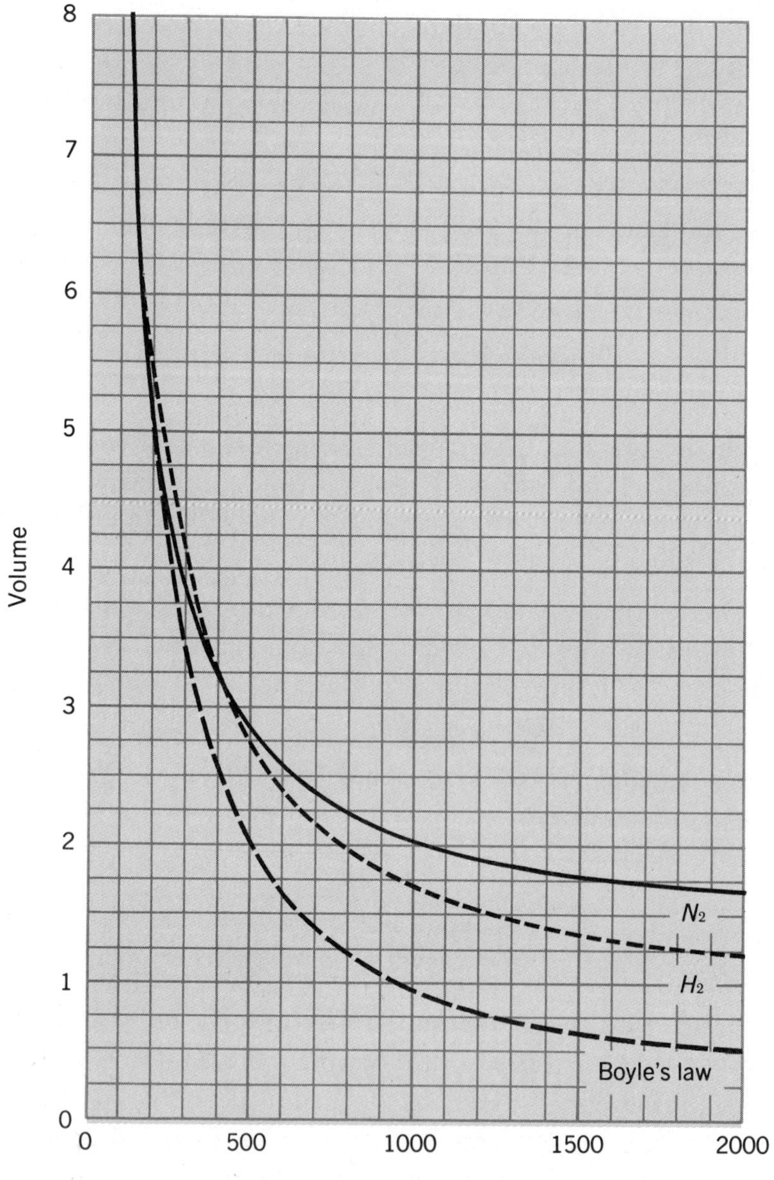

FIG. 9.16
The different compressibilities of different gases at high pressure are shown by this graph of the volume as a function of pressure for nitrogen and hydrogen at 0°C. The lowest dashed line is the curve a gas would have if it satisfied Boyle's law at high pressure.

gases differ more and more. The compressibility depends on the kind of molecule that makes up the gas.

In Sec. 9.3 we made the assumption that the bouncing of one molecule off the wall of a container is not affected by the presence of other gas molecules. This assumption is valid only at low pressures and, therefore, at low densities. Apparently at high pressures where gas molecules are close together, they affect each other's motion. Different kinds of molecules affect each other differently.

The compressibility that we observe for gases at such high pressures is about the same order of magnitude as that of liquids. Also, the density of nitrogen gas at this high pressure is nearly as great as that of water. From these observations we would predict that the compressibility of liquids (and solids as well) should depend on the kinds of molecules of which they are composed. Such, in fact, is the case. Although very high pressures are needed to compress liquids by a measurable amount, we find that the compressibility of a liquid does indeed depend upon the kind of molecules making up the liquid. For example, to compress a given volume of glycerine a given amount requires 2.5 times the pressure needed to compress the same volume of water by the same amount.

TEMPERATURE AND MOLECULAR SPEED 9.6

Let us return to the apparatus of Sec. 9.2. What do we have to do to make the sphere gas expand under constant pressure, that is, without decreasing the load on the plastic disk? You might guess that, if the spheres are made to move faster, they will strike the disk harder and more often. This will increase the pressure on the disk, and the disk will rise. As the disk rises, the spheres will have to travel longer distances between collisions with the disk. Therefore, they will not strike it so frequently as before, and thus they will exert less pressure on it. Finally, the pressure exerted by the spheres will fall to just what it was before we increased the speed of the spheres. The disk will stop rising and will remain at the higher position; thus the volume of the gas will have increased.

In the gas-model machine the speeds of the spheres depend on how fast the motor moves the platform up and down. Thus by speeding up or slowing down the motor, we can change the speeds of the spheres. Figure 9.17 shows what happens when we have the

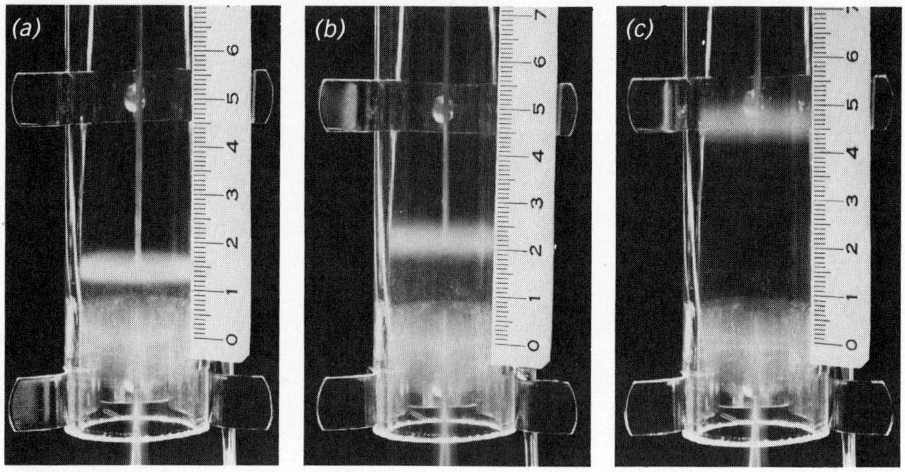

FIG. 9.17
In each of the three photographs the gas-model machine contains the same number of steel spheres, but the speed of the driving motor increases from (*a*) to (*c*).

motor running at different speeds. It is clear that, as the speeds of the spheres increase, the volume increases. Since the volume of a gas increases with temperature, this behavior of the sphere gas suggests that the temperature of a real gas is related to the speeds of the molecules. At a higher temperature the molecules move faster. At a lower temperature they move more slowly.

To check this connection between temperature and molecular speed in a real gas, we have to find a way to speed up or slow down the gas molecules and see how the temperature of the gas changes. How can this be done?

When you throw a good tennis ball at a hard wall, it bounces back with practically the same speed with which it struck the wall. On the other hand, if you hit an oncoming tennis ball with a racket moving toward it, the ball bounces back with a higher speed than it had before the collision. If many tennis balls are hit by the moving racket, they will all bounce back at higher speeds; thus their average speed will increase by collision with the moving racket.

We may try the same trick on molecules of a gas. Instead of a tennis racket we shall use a piston. Consider air in a tube closed at one end and fitted with a piston at the other end. While the piston is being pushed in, the molecules striking it will bounce back at higher speeds. In speeding up the molecules, the piston is acting like the moving tennis racket. Hence, when we compress a gas, we expect its temperature to rise. This effect can be demonstrated very convincingly by placing a little piece of cotton near the sealed end of the tube and moving the piston in with a quick push: the cotton will start burning, as shown in Fig. 9.18. (Don't try this with a test tube. The tube will burst.)

To sum up, we have seen that the temperature of a gas is related to the average speed of its molecules. When the average speed of the molecules increases, the temperature rises.

FIG. 9.18
A fire syringe. Its parts (*a*) are a protective metal outer tube, a glass inner tube with a wisp of cotton near its bottom, and a piston. (*b*) When the piston is pushed rapidly down the glass tube, the rising temperature of the gas ignites the cotton; a flash of light is seen (*c*) through the hole in the outer tube.

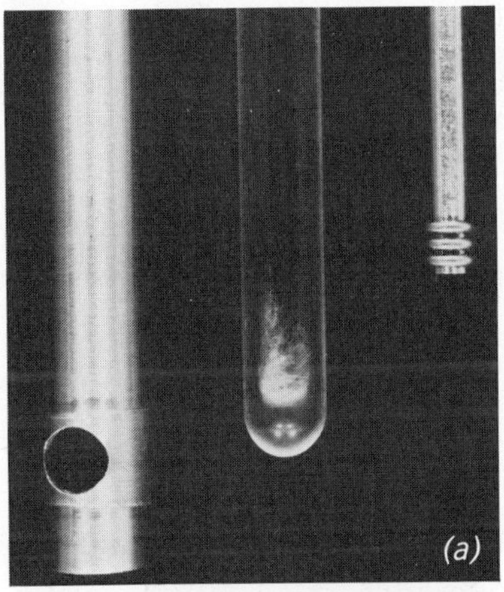

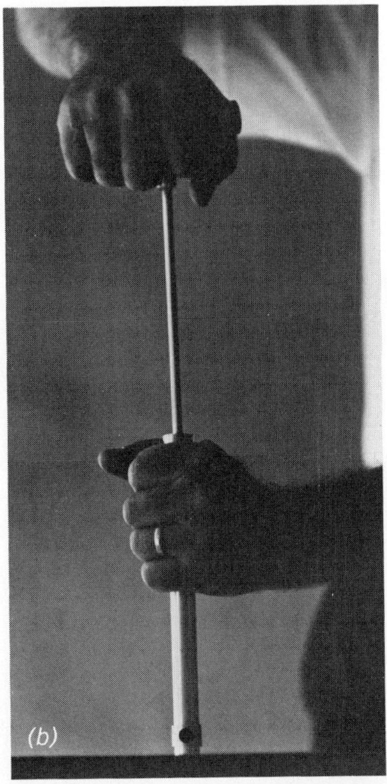

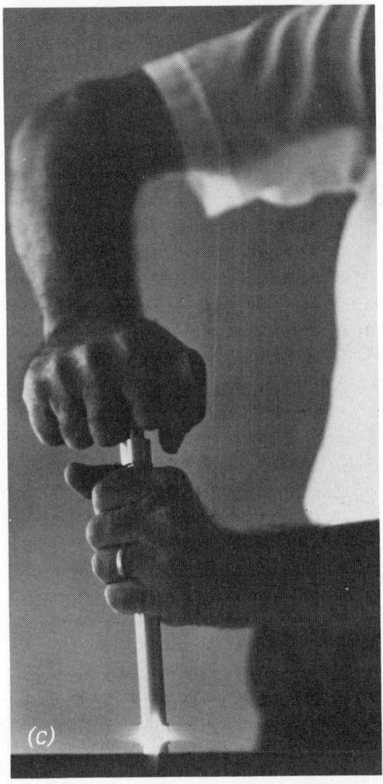

In this discussion of the thermal expansion of a gas in terms of the atomic model, we have made use of only the same two basic assumptions we used in the discussion of the compressibility of gases in the preceding section. These assumptions were (*a*) that the molecules of a gas are constantly in motion and (*b*) that the bouncing of one molecule off the wall of a container is not affected by the presence of other molecules. The assumptions do not depend on the kinds of molecule that make up the gas. Therefore, the atomic model predicts that at low pressures all gases have the same thermal expansion.

---

12 A fast-moving tennis ball strikes a racket that is moving back, away from the ball.

a) How does the speed of the ball before it hits the racket compare with the speed after it rebounds?

b) If the piston in Fig. 9.15 is pulled up, what happens to the gas molecules when they rebound?

c) What will happen to the temperature of the gas as the piston rises?

13 Does a bicycle pump heat up when you pump up a tire?

9.7 THERMAL EXPANSION OF GASES

The thermal expansion of equal volumes of three gases—air, propane, and carbon dioxide—is shown in the film loop "How Does the Thermal Expansion of Gases Compare?" The results of the experiment in the film loop are shown in Fig. 9.19. Note that the points for all three gases lie close to the same straight line. Within the accuracy of the measurements, equal initial volumes of all three gases show the same increase in volume for the same temperature change, as predicted by the atomic model. Similar experiments have been made with many other gases. The results are the same.

9.8 THERMAL EXPANSION OF LIQUIDS

When you warm up some water in a test tube, you do not notice a change in the volume. The thermal expansion of liquids is so small that we must amplify the expansion in order to observe it. This can

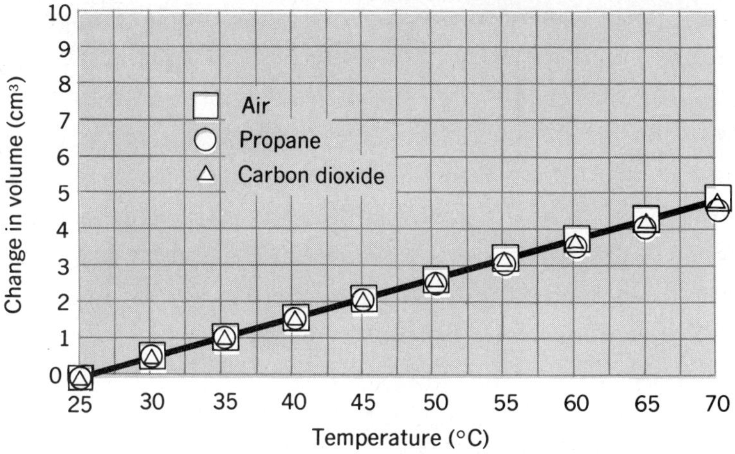

FIG. 9.19
Graphs of the thermal expansion of three different gases that had the same initial volumes. The expansion of all three is the same, independent of the kind of molecules.

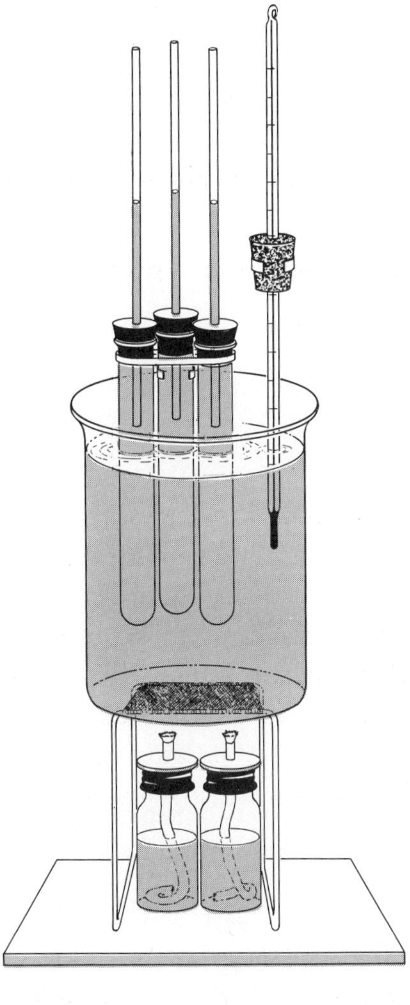

FIG. 9.20
Apparatus for comparing the thermal expansion of liquids. The test tube in the center is supported by a clamp, and the other two are fastened to the central one by a rubber band. Before heating, the height of the liquid in each of the test tubes is adjusted to be the same. Thus we start with equal volumes of liquids.

be done with the apparatus shown in Fig. 9.20. Consider one of the test tubes. Practically all the liquid is contained in the test tube, and only a small fraction is in the narrow glass tube. When the liquid expands, the small increase in volume of the liquid in the test tube pushes liquid a large distance up the narrow tube, thus amplifying the small volume change so that it can be measured. We can compare the thermal expansion of three liquids at once by putting a different one in each of the test tubes. We can measure the rise in liquid level in the narrow glass tubes when we change the temperature through a given range. This experiment has been done in the film loop "How Does the Thermal Expansion of Liquids Compare?" The results are shown in the graph in Fig. 9.21. Clearly, the thermal expansion of a liquid depends on the substance.

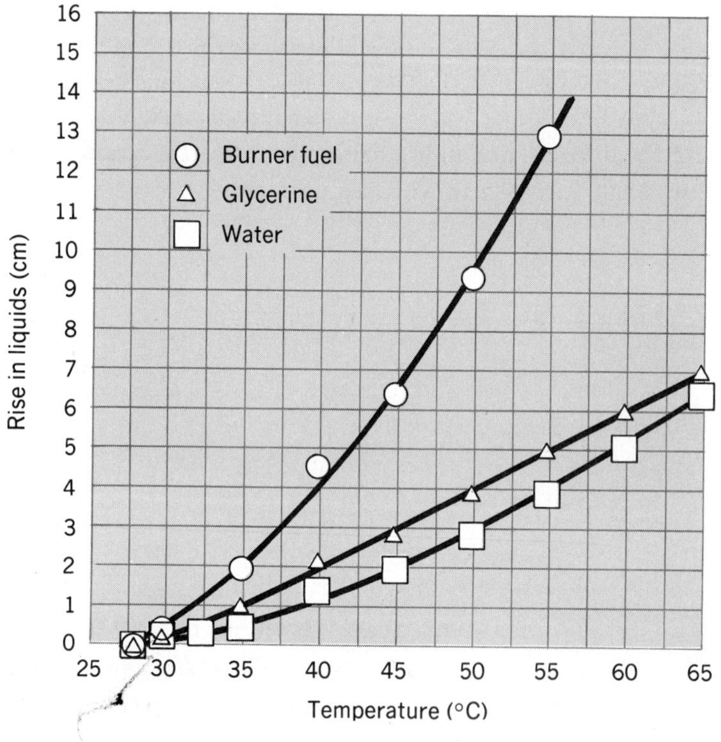

FIG. 9.21
The thermal expansion of three different liquids that had the same initial volumes. The expansion of all three is different, depending on the kind of liquid.

14 **If the two tubes in Fig. C contain the same liquid and the initial levels are the same, in which tube will the liquid rise higher as the temperature of the liquid in both tubes is raised equally?**

FIG. C For prob. 14.

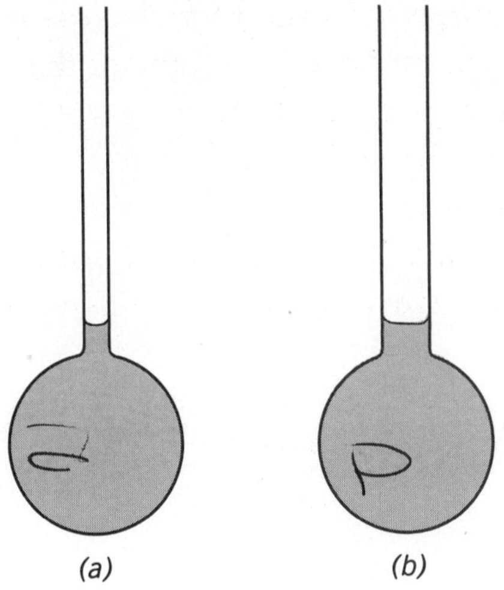

15 **One of the liquids in the apparatus in Fig. 9.20 has a volume of 33 cm³. Its tube has an internal diameter of only 0.3 cm and a volume of about 0.07 cm³ for 1 cm of its length. For a 10°C temperature rise for burner fuel, using this apparatus, the liquid rose 6.0 cm. What was the increase in volume of 1 cm³?**

EXPERIMENT THERMAL EXPANSION OF SOLIDS 9.9

From the experiment described in Sec. 9.8 we concluded that the thermal expansion of liquids differs from liquid to liquid. Is this true for solids?

In your daily life you have probably never noticed the thermal expansion of solids. It is so small that our senses are not keen enough to detect it. Sensitive apparatus must be used to amplify this very small effect so that we can observe it. Figure 9.22 shows a device that you can use to measure the thermal expansion of a tube. Fasten a metal tube in place, and connect it to a test tube half full of water. Start heating the water and set the circular dial at zero. A dish placed under the outlet of the tube will catch the condensed steam.

What does the dial read when the tube stops expanding? What is the temperature of the tube? (You can find the temperature of the escaping steam by loosely inserting a thermometer into the short rubber tube on the outlet end of the long metal tube. Make sure you hold the thermometer there long enough to get an accurate reading.)

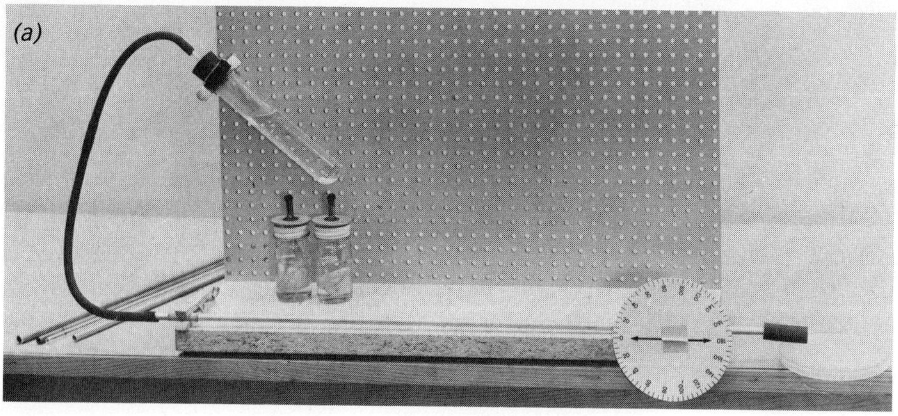

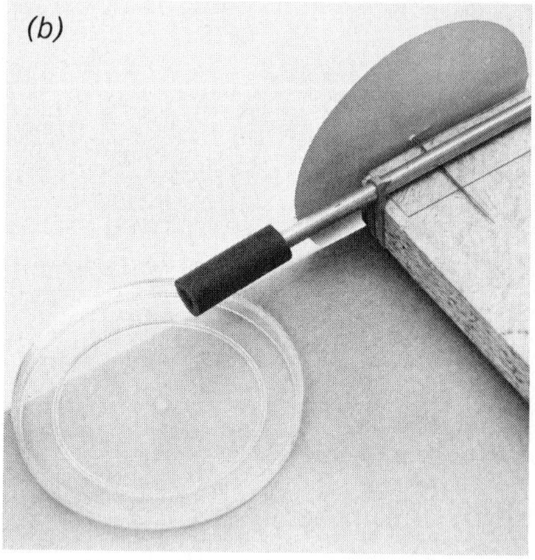

FIG. 9.22
(*a*) Apparatus for measuring the thermal expansion of a long tube. Steam from boiling water in the test tube passes through the length of the tube, heating it and causing it to expand. The left-hand end of the tube is held firmly by a clothespin, preventing this end of the long tube from moving. The right-hand end of the tube is free to move, and its expansion is amplified and measured by the rotation of the circular scale on the dial. The details of the amplifying mechanism are shown in (*b*). As the long tube expands, it moves over a needle that rolls on a smooth glass microscope slide. Attached to the needle is a dial, which amplifies the slight motion by turning through a large arc.

When the metal tube has cooled enough to allow you to handle it, replace it with a tube of the same length but of a different material.

Now measure the expansion and the temperature change of a tube of a third material of the same length. Were all the tubes heated through nearly the same temperature range? When tubes of equal length are heated from room temperature to the temperature of condensing steam, is the thermal expansion of a solid independent of the substance?

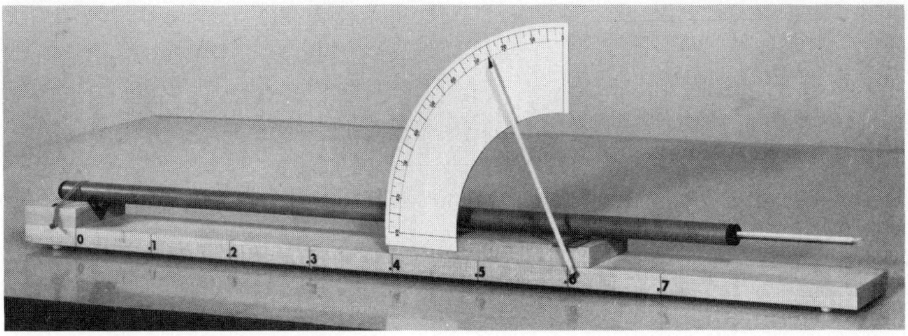

FIG. D For prob. 16.

16 **A somewhat different apparatus for studying the thermal expansion of a tube is shown in Fig. D. Here the tube is filled with hot water, and the pointer is set at zero in a vertical position.**

With this apparatus the following data were taken as the water in the tube cooled from 70.4°C to 37.8°C:

Temperature (°C)	Angle (deg)	Temperature (°C)	Angle (deg)
70.4	0	50.2	25
66.1	5	45.9	30
62.4	10	41.8	35
58.0	15	37.8	40
54.2	20		

a) Plot a graph of the angle through which the pointer turns as a function of the temperature.
b) From this form of the apparatus what information about the thermal expansion of a rod can you obtain that you were not able to get with the apparatus you actually used?

MOLECULAR MOTION IN LIQUIDS AND SOLIDS 9.10

In a cold gas the molecules move around more slowly than they do in a hot gas. If we continue to cool a gas, we expect the speed of the molecules to get less and less. If we cool it enough, it becomes a liquid. A liquid, unlike a gas, does not fill all the space available to it; the water in a glass, for example, does not spread throughout a room. Does this mean that in a liquid the molecules do not move? There is a simple though slow way to show that molecules in a liquid do move. Figure 9.23 shows a series of photographs of a graduated cylinder containing a water solution of copper sulfate and pure water. In the first photograph, the more dense, dark, copper sulfate solution is in the bottom half with pure water floating on top of it. After several days, you can see that the division between the two liquids is no longer sharp. Some of the blue copper sulfate has diffused up into the water, making it slightly colored. As you can see, after about 19 weeks the copper sulfate is spread almost evenly throughout the cylinder. This is like the behavior of bromine gas diffusing into air (Sec. 9.1), except that the copper sulfate diffuses through water much more slowly. This is what we should expect since in a liquid the molecules are much closer together than they are in a gas. Even if they move at about the same speeds as molecules in a gas, molecules in a liquid can move only a very short dis-

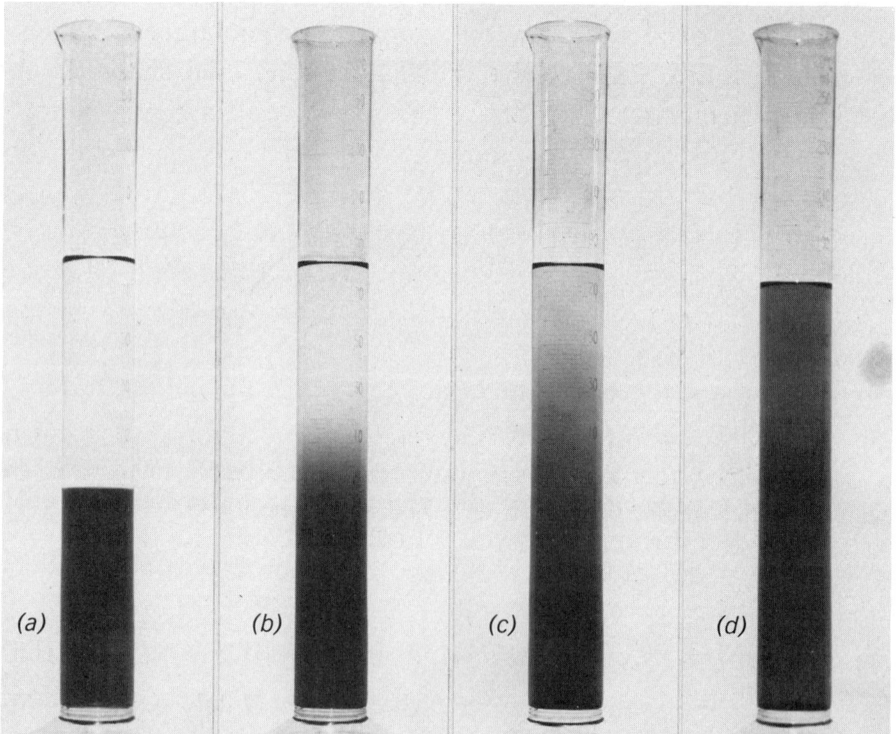

FIG. 9.23
The diffusion of copper sulfate solution into water. (*a*) At the start of the experiment the copper sulfate solution and the water form two distinct layers. (*b*) Five days later. (*c*) Thirty-seven days later. (*d*) One hundred thirty-four days later.

tance before hitting and bouncing off other molecules. Think again of the boy running fast across the crowded football field. It will take him much longer to go from one edge of the field to the other if there is a dense crowd rather than a thinly scattered one, even if he runs at the same speed in both cases.

You can convince yourself in another way that the molecules of a liquid must be in motion. Suppose we bring cold air into contact with hot water. The air will warm up. We already know that, as air warms up its molecules speed up. Hence something must be "hitting" the air molecules to speed them up; it is only the molecules in the hot water that can do that. We can apply the same argument to show that the atoms in a solid also are in motion: A hot solid heats a cold gas in contact with it.

Nevertheless, there must be a difference between the motion of the atoms (or molecules) in a liquid and in a solid since the behavior of the two kinds of materials as a whole is so different. A liquid flows easily and takes the shape of the container in which it is held. A solid, on the other hand, keeps its own shape. In the atomic picture, this means that, while the molecules in a liquid cannot move apart from one another, they can easily slide past one another. In a solid, the atoms are more or less held in their places, and the only motion possible is a kind of to-and-fro movement, or vibration. Thus the atoms on the surface of a hot solid heating a cold gas behave like the vibrating platform in our gas-model machine.

---

17 The two liquids in Fig. 9.23 were kept at room temperature during the diffusion. How would you expect the pictures to differ if the experiment had been run at a higher temperature?

EXPERIMENT GROWING SMALL CRYSTALS 9.11

You can get some idea of the arrangement of atoms in a solid by observing the growth of a crystal. Melt a pinch of salol in a watch glass over a beaker of hot water. As soon as it has melted, remove the watch glass from the beaker and observe the liquid with a magnifying lens while it cools and freezes. Describe what happens. What does the solid look like? Make a sketch showing the shape

FIG. 9.24
Crystals of alum (left) and of Rochelle salt (right).

of the crystals that form as the liquid freezes. In terms of the atomic model, what do you think happens as the liquid turns to solid?

Crystals of solid matter grow in a number of different shapes. Figure 9.24 shows two examples of different crystals. Their regular shapes suggest an atomic model for a solid in which the atoms arrange themselves in a regular, orderly pattern. The atoms become "locked" into position in orderly layers as the crystal slowly grows, giving rise to regular, smooth-faced crystals, like those shown in Fig. 9.24.

Figure 9.25 shows one simple way in which atoms can arrange themselves in a crystal. Crystals do not flow like liquids—they keep their shape. Hence we have to conclude that the motion of the atoms in the crystal is limited to vibrations about their normal, fixed positions in the crystal pattern. With this model of a solid, we can describe temperature changes in the following way: When a crystalline solid is heated, the vibrations become more rapid; when it cools, they slow down.

When a hot solid is put into contact with a cold gas, the gas warms up and the solid cools down. As you can see from Fig. 9.26, the vibrating atoms along the sides of a solid can push on gas molecules so as to speed up their motion and thus increase the tempera-

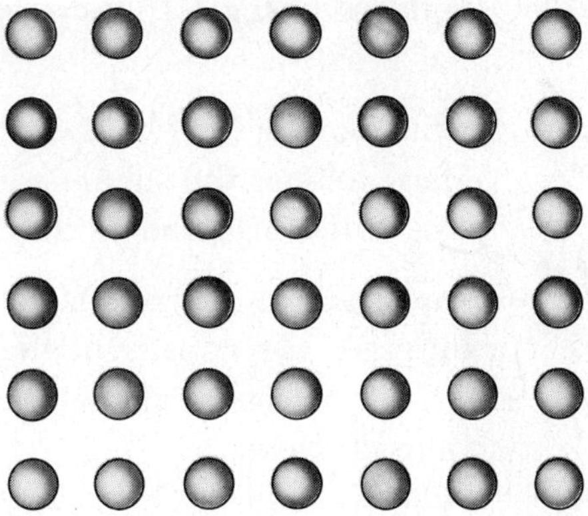

FIG. 9.25
One way in which atoms can be arranged in crystals.

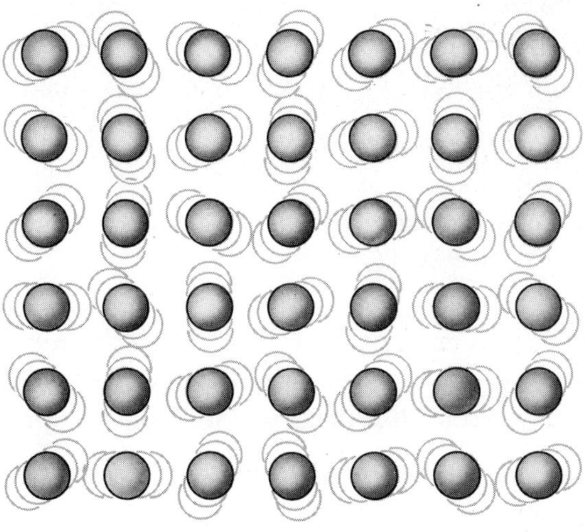

FIG. 9.26
The atoms in a solid can vibrate about their normal positions. If the atoms along the sides are struck by fast-moving gas molecules, their speed of vibration will increase.

ture of the gas. At the same time, the vibrations of the atoms in the solid decrease a bit. Conversely, the rapidly moving molecules of a hot gas or liquid in contact with a solid can speed up the vibration of the atoms in the solid, thus raising the temperature of the solid.

For Home, Desk, and Lab

18 Bromine gas has a density of 6.5×10^{-3} g/cm^3. Can this be the reason why the gas moves up through the air in the tube shown in Fig. 9.1?

19 Figure 9.2 shows that when bromine vaporizes in a vacuum, the color seems to spread immediately throughout the tube. What does this tell you about the speed of the bromine molecules?

20 Would you expect an increase in pressure on the walls of the air-filled bromine tube in Fig. 9.1 as the bromine evaporates? Why?

21 Do the moving steel balls exert any pressure on the side walls of the vertical tube of the sphere-gas machine? What changes would you make in the machine to check your answer?

22 If you run the sphere-gas machine with a long tube and without a top disk, the density of the gas is greatest near the bottom and decreases as you go up the tube. Is there a similar effect in the atmosphere of the earth?

23 As Figs. 9.5 and 9.6 show, the top disk does not stay at one place but moves rapidly up and down about a certain average position.

a) How do you account for this motion?

b) Why is such a motion not observed in the experiment described in Sec. 9.4?

24 For three or four values of volume and pressure in Fig. 9.15 calculate the volume times the corresponding pressure. How do these products compare?

25 In what different ways can you increase the pressure of a gas?

26 How is the volume of the steel-sphere gas affected by (a) motor speed, (b) number of spheres, and (c) the mass of the disk on top?

27 What two basic assumptions have we added to the atomic model developed in Chap. 8 in order to use it to describe the compressibility and thermal expansion of a gas?

28 Diesel engines do not ignite the fuel-air mixture with a spark from a spark plug as gasoline engines do. Instead, air in the cylinder is compressed by a piston. At maximum compression, fuel sprayed into the compressed air ignites and drives the piston back. How can you explain the ignition of the gas when there is no spark to ignite it?

29 a) Suppose the apparatus you saw in the film loop "How Does the Thermal Expansion of Gases Compare?" were used to study the properties of neon gas. How would you expect the experimental data to compare with that shown in Fig. 9.19?

b) Suppose the apparatus were used to study the properties of a mixture of propane and carbon dioxide. How would you expect the experimental data to compare with that shown in Fig. 9.19?

30 a) If the liquid levels in the two tubes in Fig. E are initially the same, will they be the same or different as the temperature of the liquids is lowered the same amount?

FIG. E For prob. 30.

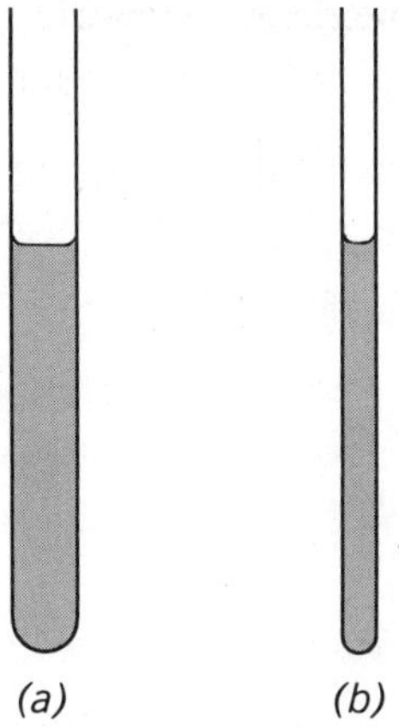

b) If the liquids in the two tubes are different, can you directly compare their thermal expansion using these two tubes?

31 The following data were obtained using an apparatus similar to that in Fig. 9.20, with water in one tube and air trapped by a drop of water in a second tube.

Temperature (°C)	Height of fluid in tube (cm)	
	Air	Water
23	0.0	0.0
25	3.9	0.4
27	7.8	0.8

What is the ratio of the change in the volume of air to the change in the volume of water for the same temperature change?

32 When heated, the glass in a thermometer bulb expands, as does the liquid that fills the bulb. Can you explain why the liquid does not go down in the tube to fill the larger volume of the bulb when the thermometer is heated?

33 Some bridges have one end firmly fixed to the support at one end of the bridge while the other end of the bridge simply rests on a roller. What do you think is the reason for this type of construction?

34 Houses are often heated by a hot-water system in which radiators and connecting pipes are completely filled with water. Such systems sometimes are connected to an open tank above the highest radiator. What do you think is the purpose of the tank?

FIG. F For prob. 35.

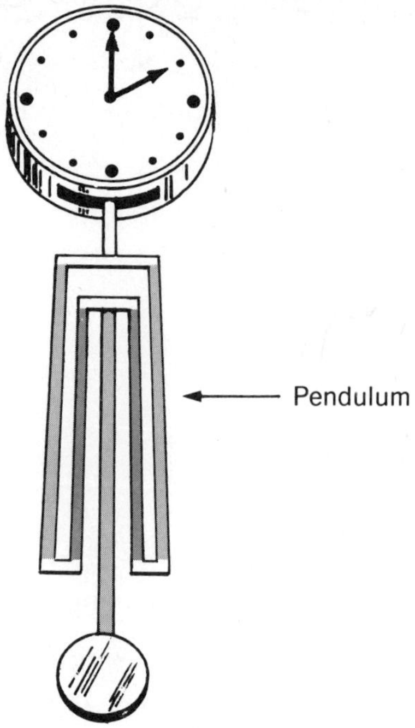

35 A pendulum clock (Fig. F) keeps time better if the length of the pendulum does not change much when there is a change in temperature. Would it be better to have the darker portion of the pendulum in Fig. F made of copper and the lighter portion made of aluminum, or vice versa?

36 If you heat one end of a short metal tube, the other end soon becomes hot. How would you account for this in terms of the atomic model?

37 Slip a thermometer into a one-hole rubber stopper, and measure its temperature. Remove the thermometer. Holding the stopper with a pencil in one end of the hole, hammer it as rapidly and as hard as you can for a minute, and then measure the temperature. What do you observe? How do you explain it in terms of the atomic model?

38 The molecules of a gas get closer together when it is compressed to very high pressures. How might you get the molecules of a gas close together without applying very high pressure to it?

Sizes and Masses of Atoms and Molecules

10

We are not able to see individual atoms or molecules when we look at a sample of water, copper wire, black ink, or any other substance. We conclude, therefore, that these tiny particles of matter are at least smaller than the smallest thing we can see. Even if we use an optical microscope to magnify a sample of an element or a compound, we cannot see the individual atoms or molecules. If we cannot see them, how can we hope to measure their sizes and masses? Clearly, we must look for some indirect way of doing this. For example, we can find out something about the size of very small particles of matter by spreading the matter out in a thin layer. Such a layer may be more than one particle thick, but it obviously cannot be thinner than the fundamental particles of which it is made.

THE THICKNESS OF A THIN LAYER 10.1

Figure 10.1(*a*) shows a sample of matter in the form of a rectangular solid. Its volume V is its length times its width times its height, or

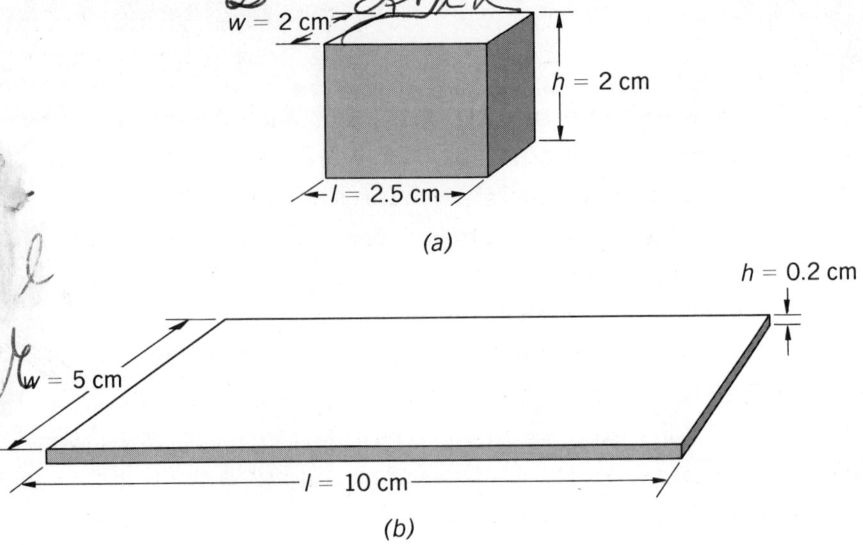

FIG. 10.1
The piece of matter in (*a*) has a volume of $lwh = 2.5\ \text{cm} \times 2.0\ \text{cm} \times 2.0\ \text{cm} = 10\ \text{cm}^3$. The flat piece of matter in (*b*) has the same volume as the matter in (*a*). The volume of the thin slab in (*b*) is $lwh = 10\ \text{cm} \times 5\ \text{cm} \times 0.2\ \text{cm} = 10\ \text{cm}^3$.

$V = lwh$. Since lw is the area of the base, we can say that the volume equals the area of the base times the height. Figure 10.1(*b*) shows the same sample of matter with the same volume as in Fig. 10.1(*a*), but with a rectangular shape that has a larger base and a much smaller height. The material has merely been formed into a different shape. Its volume is still the area of its base times its height. The material might be spread out in an even thinner layer. Its volume would still be equal to the area of its base times its height and would be unchanged.

Now suppose we have a thin rectangular sheet of some material—some metal foil, for example—whose volume we know. From measurements of its length and width we can calculate the area of its base. To find the thickness we divide the known volume by the area of the base.

---

1 **How do you know that there are more than 200 molecules in a spoonful of water?**

2† **A cube has a side of 2×10^{-7} cm. What is its volume?**

EXPERIMENT THE THICKNESS OF A THIN SHEET OF METAL 10.2

Find the thickness of a rectangular sheet of aluminum foil from its length, width, and volume. You can find the volume from the mass of the sheet and the density of the material (2.7 g/cm^3).

From your measurement of the thickness, how large can the atoms of aluminum in the foil be? How small can they be?

Metals do not flatten out into thin layers by themselves; they have to be rolled or hammered. Hammering, however, is a crude process. Even the most skilled goldsmith, making thin gold leaf for lettering on store windows, must stop when the gold leaf is about 10^{-5} cm thick. In such cases the minimum thickness is not determined by the size of the fundamental particles of gold but by the difficulty of handling such a thin sheet.

Liquids, on the other hand, tend to spread out into thin layers by themselves. If we pour water on the floor, it spreads out quickly to form a thin layer. We can calculate the thickness of this layer if we know the volume of the water we poured out and the area it covered.

3† In a small-boat harbor a careless sailor dumps overboard a quart (about 1,000 cm^3) of diesel oil. If we assume that this will spread out evenly over the surface of the water to a thickness of 10^{-4} cm, what area will be covered with oil?

4 Spherical lead shot are poured into a square tray 10 cm on a side until they completely cover the bottom. The shot are poured from the tray into a graduated cylinder, which they fill to the 20-cm^3 mark?
- **a) What is the diameter of a single shot?**
- **b) How many shot were in the tray?**
- **c) If the 20 cm^3 of shot had a mass of 130 g, what was the mass of a single shot?**

5† A tiny drop of mercury has a volume of 1.0×10^{-3} cm^3. The density of mercury is about 14 g/cm^3. What is the mass of the drop?

6† A goldsmith takes 19.3 g of gold and hammers it until he has a thin sheet of foil 100 cm in length and 100 cm in width. The density of gold is 19.3 g/cm^3.
- **a) What is the volume of gold?**
- **b) What is the area of the gold sheet?**
- **c) What is the thickness of the gold sheet?**

10.3 EXPERIMENT THE SIZE AND MASS OF AN OLEIC ACID MOLECULE

Some substances, such as oil, under proper conditions can spread out even more thinly than water. Since oil does not mix with water but floats on top of it, we often see it spread out on a water surface to a thin, rainbow-colored film. Oleic acid, a compound of hydrogen, carbon, and oxygen, does not dissolve in water but spreads out in an even thinner layer than oil does. In fact, a single drop of this liquid of the size obtained from a medicine dropper will spread so thin as to cover the entire surface of a small wading pool.

If we wish to make and measure the thinnest layer this substance will form on a small water surface, we must have some way of obtaining a measurable volume of oleic acid that is much smaller than a drop from a medicine dropper. If you know the volume of such a tiny quantity and the area of the layer it forms, the thickness is just the volume of the drop divided by the area of the layer.*

Place about one centimeter of tap water in a tray and allow it to stand for about 5 minutes so that all movement of the water has stopped. Then sprinkle on the water just enough of a fine powder to be barely visible. When a tiny amount of oleic acid is dropped on the water surface, the powder will be pushed aside, allowing us to see the oleic acid film when it spreads out across the water.

Bend a piece of fine wire into a narrow V and clean it with alcohol. When dry, dip the tip of the V into oleic acid until a very small droplet clings to the inside of the V. Take care not to immerse it more than necessary to obtain one small droplet.

To get a rough idea of the volume of the droplet it will suffice to assume that the droplet is a cube and to estimate the length of its side, using a centimeter scale and a magnifying glass.

Dip the wire tip into the water several times, until the oleic acid layer stops expanding. What is the diameter of the circular area covered by the film that is formed? If the area is only roughly circular, how would you measure its average diameter? Calculate the thickness of the layer, using your volume estimate.

In your calculation of this fantastically small thickness you used only the fact that the height of a cylinder equals its volume

*The volume of a thin cylinder or disk, like that of a rectangular sheet, is equal to the area of the base times the height, no matter how short the cylinder. In the case of a cylinder, however, the base is a circle and its area is equal to πr^2, where r is the radius of the base.

Arg

divided by the area of its base. To draw any conclusions about the size of oleic acid molecules from the thickness of the layer, we have to make an assumption: that the oleic acid spreads out until the layer is only one molecule thick. From this assumption it follows that the thickness you calculated is the height of a single molecule.

So far we have found only the height of a molecule of oleic acid. Other experiments show that oleic acid molecules are long and thin, with a height about 10 times the width of their base. They stand nearly upright on a water surface when they form a thin layer. Thus we can imagine that a tiny piece of a layer of oleic acid, one molecule thick, looks roughly like the collection of tiny rods seen in Fig. 10.2. The area of the base of each rod is then equal to (width of base)^2 as shown in Fig. 10.3.

Suppose that all the molecules "touch" one another as in Fig. 10.2. Then, total area of layer = (number of molecules) × (area of base of one molecule), or

$$\text{Number of molecules} = \frac{\text{total area of layer}}{\text{area of base of one molecule}}$$

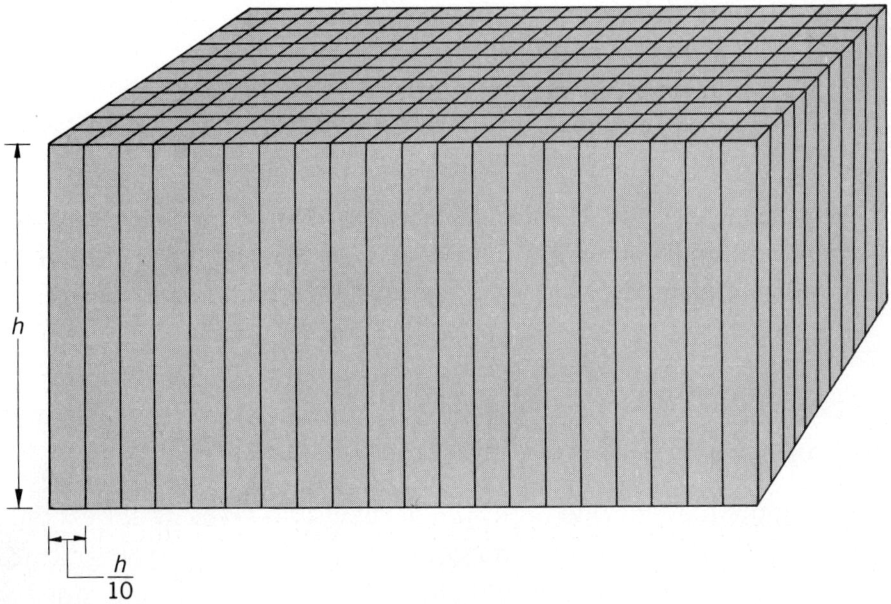

FIG. 10.2
A simplified picture of a tiny, submicroscopic piece of an oleic acid film one molecule thick.

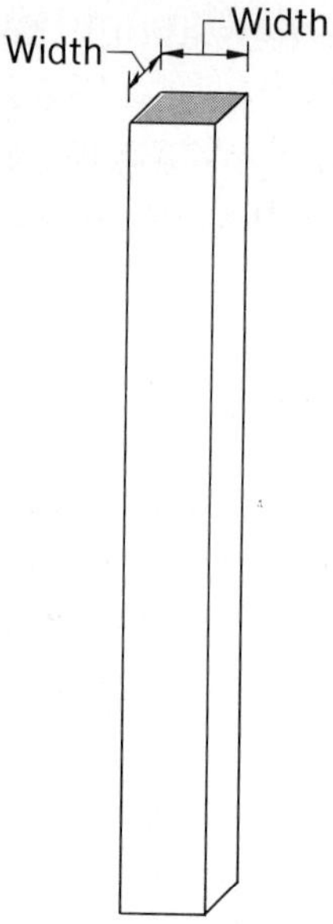

From your data for the height of one molecule and the information that the width is $\frac{1}{10}$ of the height, what is the area of the base of one molecule? What was the number of molecules in the layer? What does this tell you about the number of molecules in the droplet with which you started?

FIG. 10.3
An enlarged view of a single oleic acid molecule. The area of the base (shaded area) is equal to $(\text{width})^2$.

The density of oleic acid is about 1 g/cm^3. What was the mass of the droplet of oleic acid? Knowing the number of molecules in the droplet, you can now use the relation

$$\text{Mass of one molecule} = \frac{\text{mass of sample}}{\text{number of molecules in sample}}$$

to find the mass of a single molecule of oleic acid.

7 If 3×10^{-5} cm^3 of pure oleic acid forms an oil film with an area of 150 cm^2, how thick is the film?

8† How many oleic acid molecules occupy 1 cm^3 if the volume of one oleic acid molecule is 10^{-23} cm^3? Assume that the molecules are 10 times as high as they are wide and that there is no empty space between them.

10.4 THE MASS OF HELIUM ATOMS

Knowing the mass of one molecule of oleic acid gives us at best only a general idea of the order of magnitude of the masses of other kinds of molecules. It tells us nothing about the masses of atoms since at this point we do not know how many atoms of carbon, hydrogen, and oxygen there are in an oleic acid molecule. In this section we shall describe an experiment in which the mass of helium atoms is determined. Of course, we shall not weigh a single helium atom: no balance is sensitive enough for this purpose. Instead, we shall prepare a sample of helium and count the number of atoms in it, determine the mass of the sample, and then calculate the mass of a single atom by division:

$$\text{Mass of atom} = \frac{\text{mass of sample}}{\text{number of atoms in sample}}$$

We have chosen to find the mass of helium atoms because helium is formed in the decay of several radioactive elements and the atoms can be counted one by one with a radiation counter. For example, polonium, which is radioactive, decays only into lead and helium. The last two are stable elements and do not decay further. If we assume that each count from a radioactive sample of polonium signals the decay of one polonium atom and the formation of one lead atom and one helium atom, the clicks of the counter will give us the number of helium atoms produced. In reality, we do not have to count all the time. If the rate of decay does not change much, we have only to measure the number of counts per minute and multiply it by the number of minutes during which the helium has been collected.

We shall now describe in some detail an experiment that was performed in 1965 at the Mound Laboratory of the Monsanto Research Corporation by M. R. Hertz and C. O. Brewer, using the method we have just outlined.

A small amount of pure polonium was placed inside a quartz tube of known diameter. The air was pumped out, and the tube was then sealed (Fig. 10.4). A few days later, enough helium had already been formed in the tube to be identified by its spectrum (Fig. 10.5).

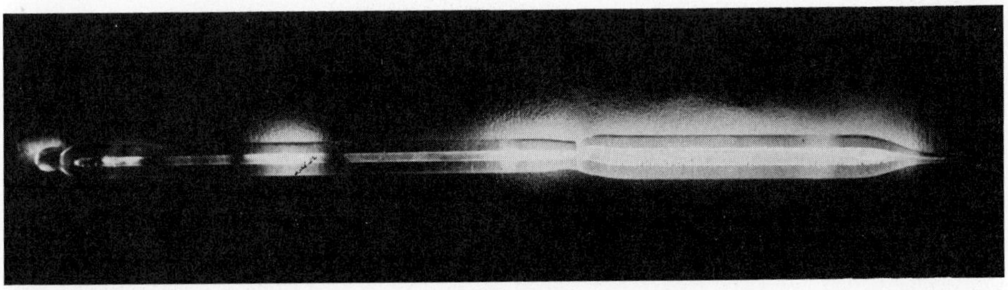

FIG. 10.4
The sealed quartz tube containing polonium. The polonium produces a blue glow, caused by the emitted helium particles when they strike the quartz. This photograph was taken in the light from the blue glow.

After about three weeks, the seal of the tube was broken under water. The water rushed into the tube until it compressed the helium to atmospheric pressure (Fig. 10.6). From the volume of the helium and its known density at atmospheric pressure, the mass of the helium was calculated.

In principle, the polonium could now be removed from the tube and placed in a counter in order to find the number of disintegrations per minute. However, a sample large enough to produce a measurable amount of helium in three weeks would give off too many helium particles per second for the counter to count. Therefore, the polonium was first dissolved completely in a large quantity

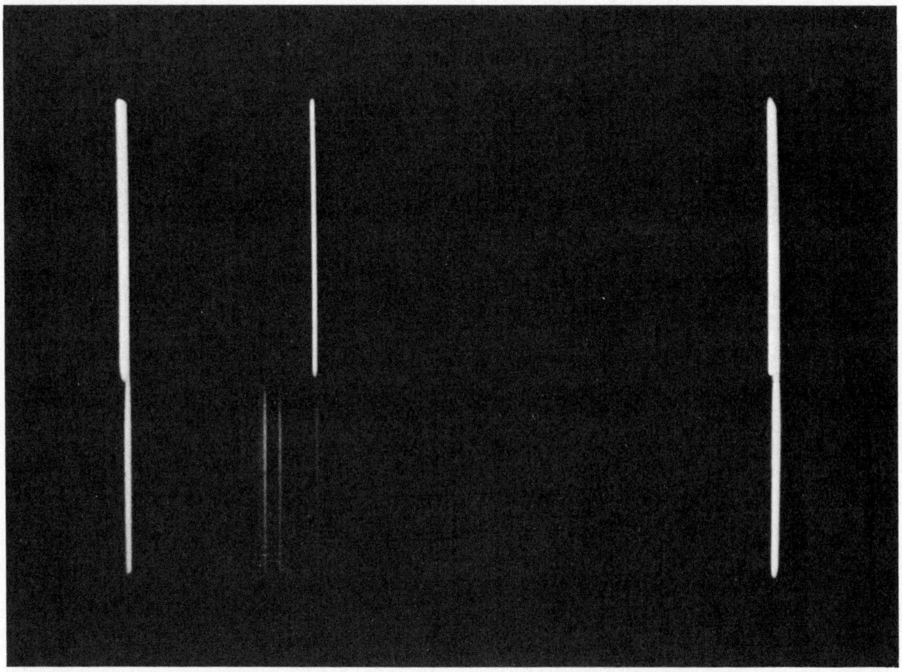

FIG. 10.5
The upper half of this photograph is the spectrum of pure helium from a tank of the gas. The lower half is the spectrum of the gas emitted by the polonium in the sealed quartz tube. The two faint lines close together in the lower half are due to polonium.

of nitric acid. The solution was further diluted with water. In both steps, the solution was thoroughly mixed to make sure the polonium was evenly distributed throughout. Finally a tiny drop of the solution was put on a little plate, the acid evaporated, and the plate placed in a counter (Fig. 10.7).

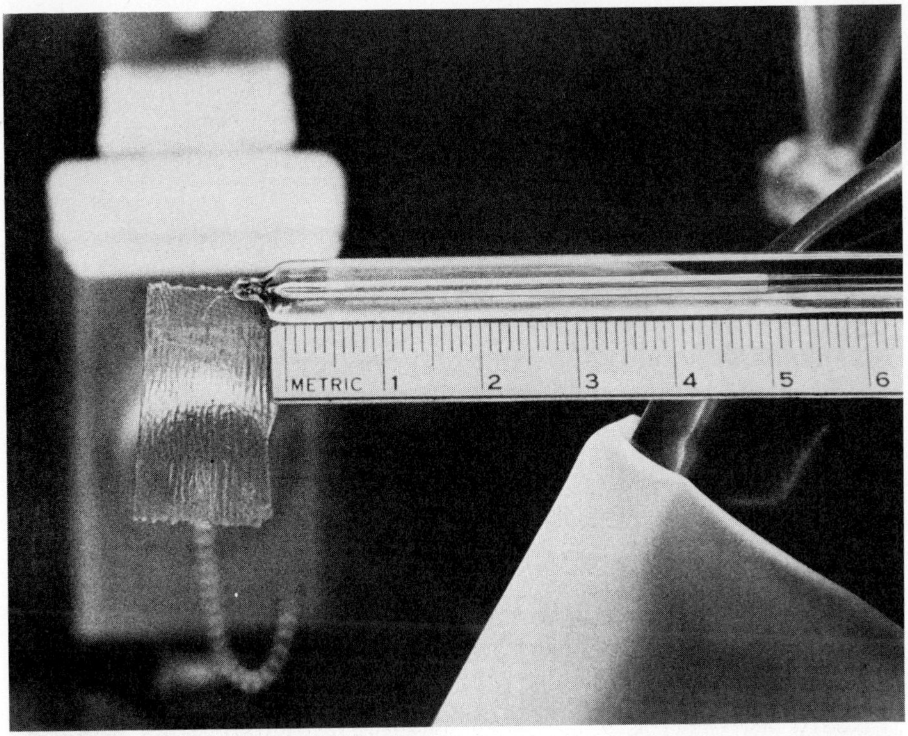

FIG. 10.6
The quartz tube containing helium after being opened under water. The length of the gas column is 5.0 cm.

FIG. 10.7
A metal plate containing polonium being removed from a radiation counter after a count.

Here is the record of the data of the experiment:

Production of Helium

On March 3, 1965, the polonium was sealed in the evacuated quartz tube. When the seal was broken on March 24, the water rose in the tube and the helium was compressed to a length of 5.0 cm in the tube.

The area of the cross section of the inside of the quartz tube, checked beforehand, was 8.1×10^{-3} cm^2. Thus the volume of helium in the tube was

$$\begin{aligned}\text{Volume} &= \text{length} \times \text{area of cross section}\\ &= 5.0 \text{ cm} \times 8.1 \times 10^{-3} \text{ cm}^2\\ &= 4.0 \times 10^{-2} \text{ cm}^3\end{aligned}$$

The density of helium at atmospheric pressure and room temperature is 1.7×10^{-4} g/cm^3. The mass of the sample of helium is

$$\begin{aligned}\text{Mass} &= \text{volume} \times \text{density}\\ &= 4.0 \times 10^{-2} \text{ cm}^3 \times 1.7 \times 10^{-4} \text{ g/cm}^3\\ &= 6.8 \times 10^{-6} \text{ g}\end{aligned}$$

Dilution

The small sample of polonium was first dissolved in 1.0×10^3 cm^3 of nitric acid. Therefore, one cubic centimeter of this solution contains $1/_{1,000}$ of the original amount of polonium.

A volume of 1.0 cm^3 of this solution was then mixed with 99 cm^3 of water. Hence 1.0 cm^3 of this dilute solution contained only 1 part in 10^5 (or 1.0×10^{-5}) of the original sample of polonium. Even this was too much to be counted!

Therefore only 1.0×10^{-3} cm^3 (one one-thousandth of a cubic centimeter!) of the very dilute polonium solution was placed on the plate to be counted. This meant that only $1.0 \times 10^{-5} \times 1.0 \times 10^{-3} = 1.0 \times 10^{-8}$, or 1 part in 100 million, of the original polonium was counted. Two trials were run with samples of this size.

Counting

Trial 1: 2.4×10^5 counts/min
Trial 2: 2.0×10^5 counts/min

This step introduces the largest experimental error in the experiment so far. The average of the two readings, 2.2×10^5 counts/min was used.

The counter used in the experiment described above does not count those helium particles which fly off into the plate, only those which fly off upward (Fig. 10.8). Thus the true disintegration rate was twice the recorded number, or 4.4×10^5 counts/min.

This large number of counts comes from only 1.0×10^{-8} of the original sample. Thus the number of disintegrations per minute of the whole sample is $1.0 \times 10^8 \times 4.4 \times 10^5 = 4.4 \times 10^{13}$ counts/min.

Now let us recall that the quartz tube containing the polonium was sealed on March 3 and opened on March 24. This is a period of 21 days. Expressed in minutes, this amounts to $21 \times 24 \times 60 = 3.0 \times 10^4$ min.

If the sample of polonium had decayed at the same rate of 4.4×10^{13} counts/min over the entire duration of the experiment, then the total number of counts would have been the product

$$4.4 \times 10^{13} \text{ counts/min} \times 3.0 \times 10^4 \text{ min} = 1.3 \times 10^{18} \text{ counts}$$

We know that this cannot be exactly the case because there are more polonium atoms at the beginning of the experiment than at the end. The number of disintegrations per minute must also be less at the end. But as we shall see in the next section, the error introduced by ignoring this change in the rate of decay is smaller than the uncertainty in the measurement of the 10^{-3}-cm^3 sample of the dilute solution. Therefore we shall use the value 1.3×10^{18} for the total number of counts during the experiment.

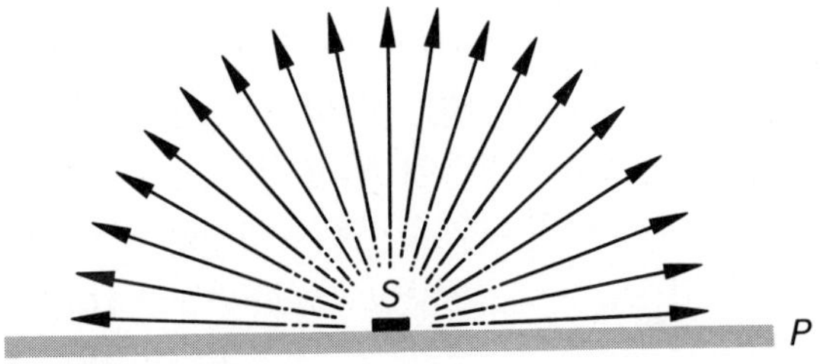

FIG. 10.8
Helium particles flying off from a thin polonium source S on a metal plate P. Those emitted in the directions shown by the arrows can be counted. Half of the helium particles are emitted downward into the plate and are not counted.

We said earlier that we shall assume that each count signals the formation of one helium atom. Thus the number of helium atoms produced by the polonium sample during the three weeks is:

$$\text{Number of atoms} = 1.3 \times 10^{18}$$

and the mass of one helium atom is

$$\text{Mass of atom} = \frac{\text{mass of sample}}{\text{number of atoms in sample}}$$

$$= \frac{6.8 \times 10^{-6}\ \text{g}}{1.3 \times 10^{18}} = 5.2 \times 10^{-24}\ \text{g}$$

---

9 If the mass of an atom of an element is 5.0×10^{-23} g, how many atoms are there in 1 g of that element?

10† A cylindrical tube with a cross-sectional area of 2 cm^2 and a height of 50 cm is filled with hydrogen. What is the volume of the hydrogen in the tube?

11 Write a brief summary of the steps followed in finding the mass of helium atoms by radioactive decay.

12† In the experiment on the mass of helium, how many lead atoms were formed? What assumptions are made to get this number?

10.5 THE MASS OF POLONIUM ATOMS

We can apply the relation

$$\text{Mass of atom} = \frac{\text{mass of sample}}{\text{number of atoms in sample}}$$

to polonium as well as to helium. We assumed that one atom of polonium disintegrates into one atom of lead and one atom of helium. Therefore, the number of atoms of polonium that disintegrated during the three weeks of the experiment equals the number of helium atoms formed. This number we calculated in the preceding section. To calculate the mass of one polonium atom we must know the mass of the polonium that disintegrated.

The decay of polonium and other radioactive elements has been studied extensively, and in all cases it has been found that the rate of decay—that is, the number of disintegrations per minute—is proportional to the amount of the element present in a

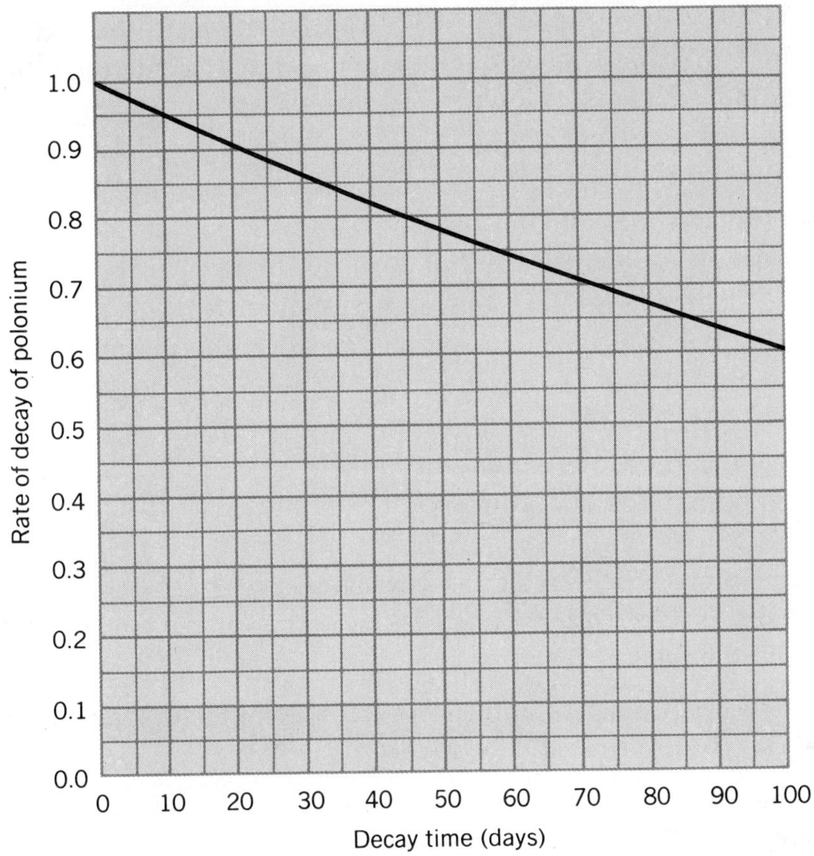

FIG. 10.9
The rate of decay of a sample containing polonium as a function of time. The rate (number of polonium atoms disintegrating per unit time) is expressed in terms of the fraction of the rate measured at zero days.

sample and does not depend on anything else. If we start, for example, with a sample of pure polonium, the rate of decay will get smaller and smaller as disintegration proceeds and fewer polonium atoms remain in the sample.

The rate of decay of polonium as a function of time is shown in Fig. 10.9. The initial rate has been chosen as unity. This enables us to use the graph for any sample of polonium. For example, you can read from the graph that the rate of decay after 21 days is 0.90 of the initial rate.*

*As you can see, in the time the tube was closed, the rate of decay of the polonium had decreased by only about 10 percent. Since we had a difference of almost 20 percent in our measurements of the counting rate of our two 10^{-8} fractions, we were justified within the accuracy of our experiment in assuming that the disintegration rate was constant and equal to 4.4×10^{13} helium atoms/min in the quartz tube.

The rate of decay in Fig. 10.9 is proportional to the fraction of the polonium remaining. For example, if only half the polonium remains, the rate of decay will be half of what it was originally. Since after 21 days the rate of decay was 0.90 of the original rate, 0.90 of the original mass of polonium still remained in the quartz tube. In other words, 0.10 of the initial sample had decayed.

All this we learn from Fig. 10.9. But to find out how much polonium decayed, we have to know the initial mass of the polonium. Although we did not mention it before, the polonium was massed before it was placed in the quartz tube (Fig. 10.10).

Massing polonium is difficult. Because of the intense and dangerous radiation, the polonium must be massed inside a small closed container. The procedure resembles the massing of a liquid. Here we quote only the final result, the mass of the container plus polonium minus the mass of the container.

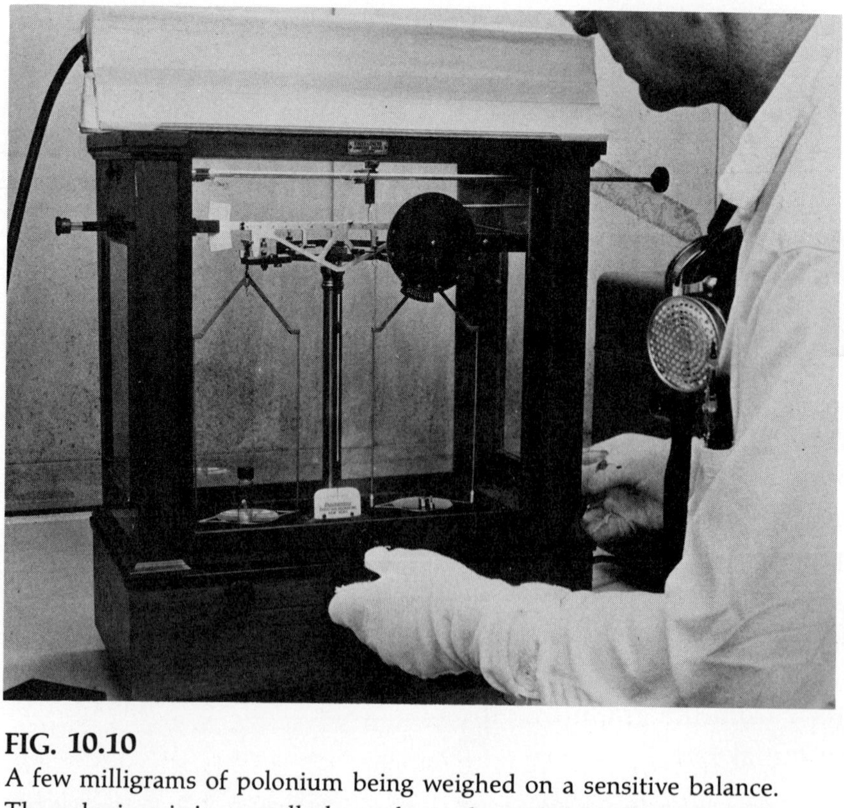

FIG. 10.10
A few milligrams of polonium being weighed on a sensitive balance. The polonium is in a small glass tube in the sealed bottle on the left-hand pan. The bottle is sealed to prevent dangerous contamination of the balance and the laboratory by the polonium.

Mass of polonium sample $= 4.5 \times 10^{-3}$ g
Mass of polonium that decayed $= 0.10 \times 4.5 \times 10^{-3}$ g
$= 4.5 \times 10^{-4}$ g

Number of atoms
of polonium that decayed $= 1.3 \times 10^{18}$

Mass of one polonium atom $= \dfrac{4.5 \times 10^{-4}}{1.3 \times 10^{18}} = 3.5 \times 10^{-22}$ g
$= 350 \times 10^{-24}$ g

Thus, writing the mass of a polonium atom to the same power of ten as we did for the helium atom, we see from this experiment that polonium atoms are about $350/5.2 = 67$ times as heavy.

The two experiments, for the determination of the number of atoms in a small sample and for the determination of the mass of atoms, are the most recent of their kind. They are by no means the first or the most accurate, but they are perhaps the most direct. A repetition of this experiment, which you may see on film, gave a similar result: 7.5×10^{-24} g for helium and 410×10^{-24} g for polonium. For higher precision, different methods are used and give a mass of 6.64×10^{-24} g for helium and 349×10^{-24} g for polonium.

Before studying this experiment, you had no way of guessing even vaguely what the mass of a single atom might be; it could be a billion times larger than 10^{-24} g, that is, about 10^{-15} g, or perhaps 10^{-33} g, a billion times smaller.

---

13 **Write a brief summary of the steps followed in finding the mass of polonium atoms by radioactive decay.**

14 **In Expt. 10.3 you determined the mass of an oleic acid molecule. How many atoms would this molecule contain if the mass of each atom equaled (a) the mass of a helium atom or (b) the mass of a polonium atom?**

THE SIZE OF POLONIUM AND HELIUM ATOMS 10.6

In Sec. 8.9 we discussed the arrangement of atoms in gases, liquids, and solids. We reasoned that the incompressibility of solids suggests that the atoms of an element in the solid state "touch" one another. Making this assumption, we can find the size of polonium atoms now that we have determined the number of polonium atoms in a sample of known mass. We shall first use the density of polonium to find the volume of the sample. Then we shall proceed to calculate the volume of one atom:

$$\text{Volume of atom} = \frac{\text{volume of sample}}{\text{number of atoms in sample}}$$

In our experiment the mass of the polonium that decayed was 4.5×10^{-4} g. The density of polonium is 9.3 g/cm^3. Thus the volume of the polonium sample that decayed was:

$$\text{Volume of sample} = \frac{4.5 \times 10^{-4}\ \text{g}}{9.3\ \text{g/cm}^3} = 4.8 \times 10^{-5}\ \text{cm}^3$$

and the volume of one polonium atom is:

$$\text{Volume of polonium atom} = \frac{4.8 \times 10^{-5}\ \text{cm}^3}{1.3 \times 10^{18}} = 3.7 \times 10^{-23}\ \text{cm}^3$$

So far we have not encountered any property of an element that would suggest a particular shape for atoms. Therefore we are free to assume a simple shape that will provide us with the easiest way to calculate a size of the atom. This shape is a cube.

The volume v of a cube of side a is given by $v = a^3$. We have just found that the volume of a polonium atom is 3.7×10^{-23} cm^3. This corresponds to a side of length 3.3×10^{-8} cm. You can check this result simply by calculating $(3.3 \times 10^{-8})^3$.

We can do a similar calculation for the size of the helium atom, but there is one difference: It is hard to get solid helium. But we can be satisfied with liquid helium since in a liquid the atoms are as close to one another as in a solid.

There were 6.8×10^{-6} g of helium produced in the experiment. The density of liquid helium (see Table 3.1) is 0.15 g/cm^3. Thus, if we had liquefied our sample of helium, it would have had a volume of:

$$\text{Volume of liquid helium} = \frac{6.8 \times 10^{-6}\ \text{g}}{0.15\ \text{g/cm}^3} = 4.5 \times 10^{-5}\ \text{cm}^3$$

This is nearly equal to the volume we found for the sample of polonium containing the same number of atoms. Therefore the calculation gives nearly the same size for both kinds of atoms. We should not take this result too literally, considering the experimental errors. Nevertheless it shows that their sizes cannot be very different.

Note that while the masses of polonium and helium atoms are very different, their sizes are about the same. This might lead you to expect that in general the sizes of atoms are fairly constant and do not depend on their mass. This is actually the case. The helium atom is the second lightest atom (only hydrogen is lighter), while polonium is one of the heaviest. Yet all atoms have a size close to 3×10^{-8} cm.

The starting point of our atomic model of matter was the idea that all matter is made up of a large number of tiny particles called atoms. Atoms can rearrange themselves in many ways but remain basically unchanged. In Chap. 8 you saw the usefulness of this idea in relating the conservation of mass to the law of constant proportions and in predicting the law of multiple proportions. In Chap. 9 you saw how the atomic model predicts the compressibility of gases. Nevertheless, one important question remained unanswered: "How small are atoms? How many atoms are there in a given sample of matter that we can see directly?" In this chapter we have answered that question.

ATOMIC MASSES AND MOLECULAR FORMULAS 10.7

In Secs. 10.4 and 10.5 we have seen a detailed demonstration of how the masses of the atoms of helium and polonium are measured. We know the masses of the atoms of all the stable elements. They were found by a variety of methods.

Although the masses of atoms vary over a wide range, they are all very small compared with the masses we are likely to put on a balance; and it would require frequent writing of factors of 10^{-24} if we expressed masses of atoms in grams. It is therefore convenient to choose the mass of the lightest known atom, that of hydrogen, as the unified atomic mass unit (u). The atomic masses of some common elements are listed in Table 10.1 in both grams and unified atomic mass units.*

*Today a unified atomic mass unit is defined differently, but the mass of a hydrogen atom, using the modern definition, is very close to 1 u. The masses given in Table 10.1 are based on the modern definition of a unified atomic mass unit.

Now that we know the masses of single atoms of the elements, we can find out how many atoms of one element there are for each atom of the other element in a compound made of two elements. Once we know this, we can write down the simplest formula for the compound. To see how this is possible, we shall consider a particular example, a liquid compound of carbon and chlorine. In this compound the ratio of the mass of chlorine to the mass of carbon has been found by experiment to be 11.8. Let us assume that the compound contains equal numbers of carbon and chlorine atoms. Then, since the mass of a chlorine atom is 35.5 u and that of a carbon atom is 12 u (see Table 10.1), the chlorine atoms would have a mass that is 35.5 u/12 u = 2.96 times that of carbon atoms. Therefore we should expect the mass of chlorine in the compound to be 2.96 times the mass of carbon. But we know that there is 11.8 times as much chlorine as carbon in the compound. Our assumption was wrong. The observed mass ratio is 11.8/2.96 = 4 times larger than the one predicted on the assumption of equal numbers of carbon and chlorine atoms in the compound. Now we can see what the answer is: There must be four times as many chlorine atoms as there are carbon atoms. Indeed 4 × 35.5 u/12 u = 11.8, the observed mass ratio. The simplest formula for the compound is, therefore, CCl_4. Its name is "carbon tetrachloride."

Carbon tetrachloride boils at 77°C, and as a gas it is made up of molecules containing one atom of carbon for every four atoms of chlorine. But on the basis of our calculation alone we cannot tell how many atoms of carbon and chlorine actually cluster together to form a molecule; the simplest formula is CCl_4, but C_2Cl_8, or C_3Cl_{12}, etc., cannot be ruled out as the correct molecular formula. Liquids and solids may not be made up of molecules at all. For these cases we are free to choose the simplest formula to describe the composition of the compound. An extension of the kind of reasoning we have used to determine the simplest formula of carbon tetrachloride would enable us to write simplest formulas for compounds containing more than two elements.

15† Assuming the smallest mass you can measure on your balance is 0.005 g, what is the smallest number of copper atoms you could mass on your balance?

Table 10.1

Atom	Mass (g)	Mass (u)	Atom	Mass (g)	Mass (u)
Aluminum	4.48×10^{-23}	27.0	Mercury	3.34×10^{-22}	201
Calcium	6.66×10^{-23}	40.1	Nitrogen	2.33×10^{-23}	14.0
Carbon	1.99×10^{-23}	12.0	Oxygen	2.66×10^{-23}	16.0
Chlorine	5.89×10^{-23}	35.5	Phosphorus	5.15×10^{-23}	31.0
Copper	1.05×10^{-22}	63.5	Potassium	6.49×10^{-23}	39.1
Gold	3.27×10^{-22}	197	Sodium	3.82×10^{-23}	23.0
Helium	6.64×10^{-24}	4.00	Strontium	1.45×10^{-22}	87.6
Hydrogen	1.66×10^{-24}	1.00	Sulfur	5.33×10^{-23}	32.1
Iodine	2.11×10^{-22}	127	Thorium	3.85×10^{-22}	232
Iron	9.27×10^{-23}	55.8	Tin	1.98×10^{-22}	119
Lead	3.44×10^{-22}	207	Uranium	3.95×10^{-22}	238
Lithium	1.15×10^{-23}	6.94	Zinc	1.09×10^{-22}	65.4

16 **Use Table 10.1 to find the mass in unified atomic mass units of a molecule of phosphorus pentoxide (P_2O_5).**

17 **In Sec. 6.2 you found that the mass ratio of hydrogen to oxygen in water was 0.13 = $\frac{1}{8}$. Using Table 10.1 find the simplest formula for water.**

18 **In Table 8.3, the ratio of the mass of chlorine to iron in iron dichloride is given as 127/100. Use information from Table 10.1 to determine the simplest formula for iron dichloride.**

For Home, Desk, and Lab

19 **A rectangular object 3.0 cm × 4.0 cm × 5.0 cm is made of many tiny cubes, each 1.0×10^{-2} cm on a side. How many cubes does the object contain?**

20 **The diameter of a tennis ball is about 0.07 m, and the dimensions of a tennis court are 15 m × 30 m. How many tennis balls will be required to cover the court?**

21 **If the molecules of the oleic acid monolayer you made could be placed end to end in a line, about how long would it be?**

22 If a 10^{-3}-g sample of radium gives a count of 4×10^7 counts/min, how much radium would give 100 counts/min?

23 Suppose you buy a 2-kg bag of dried beans and you find it has been contaminated with small stones. How would you go about finding the approximate number of the stones in the bag without separating all the stones in the bag from the beans? What assumptions have you made?

24 A cubic millimeter (10^{-3} cm^3) of blood is found to contain about 5×10^6 red blood cells. An adult human body contains about 5×10^3 cm^3 of blood. About how many red blood cells are there in an adult human body?

25 Some polonium is dissolved in 1,000 cm^3 of nitric acid, and a 0.01-cm^3 sample of the solution is counted. Then the number of disintegrations per minute is found to be 3×10^3. How many disintegrations per minute occurred in the original solution?

26 If 10^{18} atoms of polonium disintegrate to produce lead and 10^{-5} g of helium, what is the mass of a helium atom?

27 Use Fig. 10.9 to determine the fraction of a polonium sample that remains after 45 days. What fraction of the polonium decayed during this time?

28 a) What fraction of a sample of pure polonium will decay in 100 days? (See Fig. 10.9.)
b) If a counter initially records 5×10^4 counts/min for the sample, what would you expect it to record after 100 days?

29 What would you get for the volume of one atom of helium if you calculated it from the equation below?

$$\text{Volume of atom} = \frac{\text{volume of gas sample}}{\text{number of atoms in gas sample}}$$

30 A penny is about 1 mm thick. About how many layers of copper atoms does it contain?

31 What is the ratio of the mass of carbon to the mass of oxygen in carbon dioxide (CO_2)?

32 Using the atomic masses of copper and chlorine in Table 10.1 return to the data you obtained from the experiment with two chlorides of copper in Chap. 8, and determine the simplest formulas for the two compounds.

33 Could a sample containing only sodium and oxygen have a molecular formula NaO if the ratio

$$\frac{\text{Mass sodium}}{\text{Mass oxygen}}$$

in the compound is 2.9? Explain. (See Table 10.1.)

34 What ratio of combining masses would you expect in a compound containing (a) one atom of lead to every atom of oxygen or (b) one atom of lead to every two atoms of oxygen?

Epilogue

As this course comes to an end, you may ask yourself, "What have I learned this year in science?" We hope you will think of several things, some specific and some of a more general nature. Look back at the questions raised in Chap. 1. To some of these you still do not know the answers ("Is there a connection between brittleness and electrical conductivity?"). To others, which may have appeared to you equally perplexing ("What do you mean by a substance?") or even ridiculous ("When you heat something, does its temperature always rise?"), you now have answers based on your own experience in investigating matter.

The purpose of Chap. 1 was to get things started. In the following chapters we tried to familiarize you with some of the basic facts and ideas of physical science, the evidence for the facts, and the usefulness of the ideas. Contrary to what you may have expected, science does not deal with absolute truths. The specific facts we find in the laboratory, such as masses, lengths, melting points, and solubilities, are all subject to the limitations of our measurements. The useful generalizations based on these measurements, the laws—such as the law of conservation of mass and Boyle's law—also have their limitations.

If this is the case in science, where we can perform experiments under controlled conditions and repeat them as many times as we wish to assure ourselves of the results, how careful must you be about the facts and generalizations you encounter in your daily life? Do you ask for evidence to support what you read and hear? If your introduction to science has made you a more critical reader, a more careful observer, and a sharper thinker, your work during the year was worthwhile.

Appendix

Answers to Problems Marked With a Dagger (†)

Chapter 2

1. (a) 8 cubes
   (b) 27 cubes
   (c) 8 cm^3; 27 cm^3
5. (a) 0.1 cm^3
   (b) 0.2 cm^3
9. (a) 10 cm^3
   (b) 20 cm^3
   (c) 20 cm^3
   (d) 30 cm^3
   (e) 0.60
14. 18.325 g
19. Boil away or evaporate the water. The mass of salt recovered would be the same as at the beginning.
20. No.
21. (a) 2.956 g
    (b) 0.054 g
    (c) About 2 percent

Chapter 3

2. (a) A
   (b) C
   (c) B
7. Measure length, width, thickness, and mass of the rectangular blocks. Multiply length by width by thickness to get volume. Then divide the mass by the volume.
9. 1.7 g/cm^3
11. (a) 10.5 g/cm^3
    (b) 2.1 g/cm^3
    (c) 0.82 g/cm^3
13. (a) 0.50 g
    (b) 1.1×10^{-3} g/cm^3
21. (a) 40 g
    (b) 50 g
22. (a) Gas
    (b) Solid
    (c) Solid or liquid
    (d) Solid or liquid
    (e) Gas

Chapter 4

10. From Fig. 4.3: 47°C
12. 212 g
15. The solids are not the same.
20. The solubility of oxygen in water decreases as the temperature of the water rises.
23. No.

Chapter 5

1. They must have different boiling points.
5. By distilling the seawater.
6. By using two sieves. One sieve allows pebbles and sand to go through but keeps back the larger stones. The other sieve

allows only the sand to go through, while keeping back the pebbles.
7. They must differ greatly in solubility.
16. The temperature of the ice (0°C) is above the boiling point of liquid nitrogen (−196°C) as given in Table 5.2.

Chapter 6

1. (a) Sodium chlorate and sodium chloride
   (b) By fractional crystallization
3. (a) 180 g
   (b) 180 g
4. No
6. (a) 18 cm^3
   (b) 2.4×10^4 cm^3
9. Tube I: 25 cm^3
   Tube II: 25 cm^3
   Tube III: 0 cm^3
11. First package / Second Package
   (a) 1.5 / 1.5
   (b) 0.6 / 0.6
13. Ratios 2, 3, 4, 6, 7
15. (a) No. (See Table 5.1)
   (b) No. Gasoline will vary from pump to pump.

Chapter 7

6. Magnésie — Magnesium, oxygen
   Baryte — Barium, oxygen
   Alumine — Aluminum, oxygen
   Silice — Silicon, oxygen
8. Minerals in the soup or milk contain metals that give characteristic colors to the flame. Since sodium compounds are almost universally present in foods, the yellow color of the sodium flame is almost always observed in cases like this.
11. Y
12. Yes
13. (a) Nothing.
   (b) The sample remained the longest on the most intensely exposed area.

Chapter 8

1. It must suggest at least one new experiment and correctly predict the results.
9. 1,300 g
15. $1.5 = 3/2$
16. $M_2/M_1 = 2.00$
   $M_3/M_1 = 3.00$
   $M_4/M_1 = 4.00$
   Yes.

Chapter 9

2. The spreading of bromine gas when a drop of bromine evaporates; the appearance of water drops on cold window panes from a container of hot water; and the odor of ammonia spreading throughout a room are all bits of evidence suggesting that atoms are in motion.
3. The density will be doubled.
7. (a) 3P
   (b) $4/3$P
10. (a) 2P
   (b) 10P

Chapter 10

2. 8×10^{-21} cm^3
3. 10^7 cm^2
5. 1.4×10^{-2} g
6. (a) 1.0 cm^3
   (b) 1.0×10^4 cm^2
   (c) 1.0×10^{-4} cm
8. 10^{23}
10. 100 cm^3
12. 1.3×10^{18}. We assume that the disintegration of one polonium atom produced one helium atom and one lead atom.
15. About 5×10^{19}

Acknowledgments

The following members of the Educational Services Incorporated staff in addition to myself were involved in the development of the course: John B. Coulter, on leave from Pakuranga College, Howick, New Zealand; Judson B. Cross; John H. Dodge; Robert W. Estin, on leave from Roosevelt University, Chicago, Illinois; Malcolm H. Forbes; Ervin H. Hoffart; Gerardo Melcher, on leave from the University of Chile, Santiago, Chile; Harold A. Pratt, on leave from Jefferson County Public Schools, Lakewood, Colorado; Louis E. Smith, on leave from San Diego State College, San Diego, California; Darrel W. Tomer, on leave from Hanford Union High School, Hanford, California; and James A. Walter.

For the summer of 1963, we were joined by Elmer L. Galley, Mott Program of the Flint Public Schools, Flint, Michigan; Edward A. Shore, The Putney School, Putney, Vermont; and Byron L. Youtz, Reed College, Portland, Oregon.

Later on, considerable time was devoted to this project by others who joined us during the summers or consulted on a part-time basis throughout the following years: Gilbert H. Daenzer, Lutheran High School Central, St. Louis, Missouri; Thomas J. Dillon, Concord-Carlisle High School, Concord, Massachusetts; Winslow Durgin, Xavier High School, Concord, Massachusetts; Alan Holden, Bell Telephone Laboratories, Murray Hill, New Jersey; Robert Gardner, Salisbury School, Salisbury, Connecticut; Father John Kerdiejus, S. J., Xavier High School, Concord, Massachusetts; Herman H. Kirkpatrick, Roosevelt High School, Des Moines, Iowa; Elisabeth Lincoln, Dana Hall School, Wellesley, Massachusetts; John V. Manuelian, Warren Junior High School, Newton, Massachusetts; John N. Meade, Newman Junior High School, Needham, Massachusetts; Paul Meunier, Marshfield High School, Marshfield, Massachusetts; Father Patrick Nowlan, O.S.A., Monsignor Bonner High School, Drexel Hill, Pennsylvania; Frank Oppenheimer,

University of Colorado, Boulder, Colorado; Charles M. Shull, Jr., Colorado School of Mines, Golden, Colorado; Malcolm K. Smith, Massachusetts Institute of Technology, Cambridge, Massachusetts; Moddie D. Taylor, Howard University, Washington, D.C.; Carol A. Wallbank, Dighton-Rehoboth Regional High School, Rehoboth, Massachusetts; Richard Whitney, Roxbury Latin School, West Roxbury, Massachusetts; Marvin Williams, Bell Junior High School, Golden, Colorado; M. Kent Wilson, Tufts University, Medford, Massachusetts; and Carl Worster, Belmont Junior High School, Lakewood, Colorado.

I also wish to acknowledge the invaluable services of George D. Cope and Joan E. Hamblin in photography; R. Paul Larkin as art director for the preliminary edition; Barbara Griffin, Nancy Nelson, and Gertrude Rogers in organization of feedback from the pilot schools; Nathaniel C. Burwash and John W. DeRoy in apparatus construction and design; Benjamin T. Richards for production; and Andrea G. Julian for editorial assistance. Much of the administrative work was done by Geraldine Kline.

Throughout the entire project I benefited from the advice and criticism of M. Kent Wilson. Valuable assistance in coordinating various group efforts in the summers of 1963 and 1966 was provided by Byron L. Youtz. In editing this edition of the course, I was specially aided by Judson B. Cross and by Harold A. Pratt, who was responsible for the group summarizing the feedback.

I wish to thank the editorial and art staff of the Educational Book Division of Prentice-Hall, Inc., for their help in preparing the final form of this edition.

Constant sources of encouragement and constructive criticism were the pilot teachers, who voluntarily spent many extra hours relating to us their classroom experience. Without them the course could not have been developed to this point.

The initial stage of the Introductory Physical Science Program was funded by Educational Services Incorporated. Since then it has been supported by a grant from the National Science Foundation. This financial support is gratefully acknowledged.

Uri Haber-Schaim

March 1967

Index

Index

A

Acids, 59, 64, 98-100, 104, 112
Air:
- components of, 81
- inflammable, 61
- liquefaction of, 81-82

Alcohol, wood and grain, 56-57
Alcohol vini, 57
Alum, 178
Ammonia gas, 62
- solubility of, 63-64

Asphalt, 73
Atomic distances, 146
Atomic mass unit (u), 199
Atomic masses, 199-201
Atomic model, 149
- of crystals, 179
- of a gas, 154-165
- of matter, 132-134, 137-138
- and molecular motion, 154
- and radioactive elements, 148-149

Atoms:
- first concept of, 1
- in gases, 146
- in solids, 145
- vibrations of, 178-179

B

Balance:
- calibration of, 19
- earliest, 16
- equal-arm, 16-19
- precision of, 20
- and use of a rider, 19

Battery, invention of, 112
Becquerel, Henri, 117
Beqa, 17
Berthollet, Claude L., 101
Black boxes, 130-132
Boiling, uneven, 35
Boiling point(s), 35-37
- tables of, 45, 57, 59, 71, 81

Boyle, Robert, 165
Boyle's law, 165
- deviation from, 166

Brewer, C.O., 189
Bromine, diffusion of, 152-153
Bunsen, Robert W. , 117

C

Calcium:
- discovery of, 112
- spectral lines of, 115

Calcium carbonate, 88
Calibration:
- of a balance, 19
- of a thermometer, 83

Candle wax, cooling curve, 31-32
Cap rock, 73
Carbon dioxide, 61-62
- dissolved in water, 63

Carbon tetrachloride, 200
Centimeter, 9
Cesium, discovery of, 117
Chlorides of copper, 139-140
Chlorine, discovery of, 112-113
Chromatography, 80
Citric acid, properties of, 58
Cloud chamber, 125-126
Complete reactions, 104-105
Composition, constant, 134-136
Compounds, 90-95
- and the atomic model, 132-134
- defined, 94
- fastener and ring experiment, 134-136
- and mixtures, 85-87
- molecules of, 144-146
- and spectral analysis, 117

Compressibility, 15-16, 43, 145
- of gases, 146, 148, 160-167
  - at high pressures, 166-167
- of liquids, 167
- of solids, 145, 148

Concentration, 50
- of saturated solutions, 51-52

Condenser, 73
Conservation of mass, 23-24, 101, 133
Constant composition, 134-136
Constant proportions, law of, 101-102, 133
Cooling curve, 31-32
Copper:
- chlorides of, 139-140
- compounds of, 138-140
- heated with sulfur, 22
- oxidation of, 102-103
- solubility in acids, 65

Copper oxide, reduction of, 104
Copper sulfate, diffusion of, 176-177
Copper sulfide, 101-102
Crude oil, 72
Crystallization, 53
- fractional, 79

Crystals, 177-179
Cubic centimeter, 9

D

Dalton, John, 142
Davy, Sir Humphry, 112-113
Decay, radioactive, 121-123, 148, 189
- rate of, 194-195

Decomposition:
- of lime, 111-112
- of potash, 112
- radioactive, 121-123, 148, 189, 194-195
- of sodium chlorate, 90-93
- of water, 93-95

Democritus, 1
Densities, tables of, 43, 45, 57, 59, 81
Density, 36-37
- compared to concentration, 50
- defined, 37
- of a gas, 40-42, 154-156
- of liquids, 39-40
- range of, 43
- of solids, 38-39

Diesel engines, 180
Diffusion, 151-154
- defined, 154
- in liquids, 176-177

Displacement of water, 13-15
Distillation:
- fractional, 67-69, 85
- of petroleum, 70-71
- of wine, 57
- of wood, 3-6, 56-57

Distillation fractions, 67-69
- of petroleum, 70-71

Dry Ice, 62

E

Earth's crust, cross section, 72
Electrolysis, 93-94
- and discovery of elements, 112

Elements, 107-127
- atomic masses of, 199-201

Elements (continued):
and atomic model, 132-133
defined, 108
discovered through electrolysis, 112
flame tests of, 113
Lavoisier's list, 109
modern list, 110
radioactive, 118-120, 122, 148-149
spectra of, 113-114
Equal-arm balance, 16-19
Ethanol, properties of, 57
Expansion, thermal, 170-172
Exponent, 24, 43

F

Fasteners and rings, 134-136
Fermentation, 56
Filtering, 74
Filtrate, 75
Flame tests, 113
Formulas, molecular, 199-200
Fractional crystallization, 79
Fractional distillation, 67-69, 85
of petroleum, 70-71
Freezing, 177-178
Freezing point, 30-33
Frozen solutions, 143
Fuels:
alcohols, used as, 57
oils, 70-73

G

Gas thermometer, 83
Gases:
compressibility of, 146, 148, 160-165
cooling, 82
density of, 40-42, 154-156
diffusion of, 151-154
at high pressures, 165-167
marsh, 61
mass of, 23
measuring volume of, 15-16
molecular structure of, 145-148
natural, 73
pressure of, 43, 76-77, 154-156
pressure-volume relation, 157-165
solubility of, 62-63
thermal expansion of, 170
Gas-model machine, 154-155, 167-168
Gay-Lussite, 88
Geiger counter, 120-121, 125
Glass-wool plug, 90-91
Graduate, 10
Graduated cylinder, 9, 13
Grain alcohol:
distillation of, 56-57
melting point, 46
Gram, 17
Grating spectroscope, 114
Green vitriol, 59, 64

H

Heating of radioactive elements, 122
Helium:
discovery of, 116
mass of atoms of, 188-194
from radioactive decomposition, 122
size of atoms of, 147, 197-199
spectrum of, 190
used in gas thermometers, 83
Hertz, M. R., 189
High pressures, 165-167
"How Does the Thermal Expansion of Liquids Compare?" 170, 172, 180
Hydrochloric acid, 64, 98, 104, 112
Hydrogen:
a component of water, 93-97
discovery of, 60-61

I

Ice:
Dry, 62
melting of, 21
Incomplete reactions, 104-105
Inflammable air, 61
International standard of mass, 17
International standard of volume, 9

K

Kilogram, 17

L

Lavoisier, Antoine L., 109, 111
Lavoisier's list of elements, 109
Laws of nature, 25, 101-102
Lead:
oxides of, 144
from radioactive decomposition, 122
Lime, decomposition of, 109-110
Limewater test, 60
Liquefaction of air, 81-82
Liquid air machines, 83
Liquid thermometer, 83
Liquids:
compressibility, 167
density of, 39-40
diffusion in, 176-177
freezing of, 177-178
molecular motion in, 175-177
molecular structure of, 145-146
thin layers of, 185
volume of, 9-10
Lithium, spectral lines of, 115

M

Magnesium, dissolved in sulfuric acid, 59
Marsh gas, 61
Mass, 16-18, 29
atomic, 199-201
conservation of, 23-24, 133
experiments with, 21-23
of gas, 23
of helium atoms, 188-194
international standard of, 17
of matter, 16-17
of polonium atoms, 194-197
of unit volume, 36
units of, 17
Mass ratios, 140-143
Matter:
atomic model of, 132-134, 137-138
and conservation of mass, 23-24
mass of, 16-17
particles of, 129-130
volume of, 15-16
Melting points, 30-33
tables of, 45, 57, 71, 81
Mercuric oxide, 87, 90

Metals, flattened into thin
layers, 185
Meter, 9
Methane, 61
Methanol:
properties of, 57
as a solvent, 58
Metric ruler, 11
Metric system, 9, 17
Micro-melting point, 33
Milligram, 19
Milliliter, 9
Mixtures:
and the atomic model, 136
and compounds, 85-87
of gases, 80-82
and pure substances, 84-87
of solids, 74-77
of solids and liquids, 73-74
of soluble solids, 76-77
Models, 129-131
Molecular formulas, 199-200
Molecular motion, 151-154
and the atomic model, 154
in crystals, 178
and diffusion, 151-154
in liquids and solids, 175-177
and temperature,
167-170, 178
vibrations, 177-178
Molecules, 144-148
definition of, 146
size of, 186-188
Multiple proportion, law of,
140-143
Muriatic acid, *see*
Hydrochloric acid

N

Nature, laws of, 25, 101-102
Negative exponent, 43
Nitric acid, 64
Nitrogen, 80-82
oxides of, 141
Nonporous rock, 72-73

O

Oil by-products, 70-73
Oleic acid molecules, 186-188
Oxidation of copper, 102-103
Oxides:
of lead, 144
of nitrogen, 141
Oxygen, 80-82
a component of water, 93-97
Oxymuratic acid, 109, 112

P

Paper chromatography, 80
Particle model of matter, 129-130
Per, defined, 36
Petroleum, 70-73
Photographic exposure by
radioactivity, 118-120,
123-125
Polonium:
danger in massing, 196
decay rate of, 194-195
mass of atoms of, 194-197
and photographic film,
123-125
radioactive decomposition
of, 121-122
size of atoms of, 197,199
spectrum of, 190
Potash, decomposition of, 112
Potassium, discovery of, 112
Potassium sulfate, solubility of,
53-54
Powers-of-10 notation, 24, 43
Precipitate, 53, 87
Prediction from the atomic model,
137-140
Pressure of a gas, 43, 76-77,
154-165
Properties of a substance, 2-3,
29-30, 45
and fractional distillation,
85
Proportions:
constant, 101-102, 133
multiple, 140-143
Proust, Joseph L., 101
Pure substances, 84-87

R

Radiation counter, 189, 191
Radioactive decay, *see*
Radioactive decomposition
Radioactive decomposition,
121-123, 148, 189
defined, 122
rate of, 194-195
Radioactive elements, 118-120
and the atomic model,
148-149
effect on photographic film,
118-120
heating of, 122
"Radioactive Substances," 118,
121, 126
Rainbow, 114
Rider, use of, 19
Rochelle salt, 178
Rounding off, 36

S

Salt:
boiling point of, 46
mass of, 21
solubility of, 14-15, 21, 54, 92
Saturated solution, 50
Saturation, 50
Scales, reading, 11-13
Sediments, 72
Shellfish, and carbon dioxide, 63
Significant digits, 37
Soda, decomposition of, 112
Sodium:
discovery of, 112
spectral lines of, 114-115
Sodium carbonate, 88
Sodium chlorate, 90-93
Solids:
compressibility of, 145
crystals of, 178
density of, 38-39
direct separation from
liquids, 73-74
dissolved in water, 49-50
heat transfers with gas,
178-179
mixture of, 74-75
molecular motion in, 177
molecular structure of,
145-146
thermal expansion of,
173-175
Solubility, 49-65
of ammonia gas, 63-64
effect of temperature on,
53-54
and filtering, 74
of gases, 62-63
measurement of, 51-52
of salt, 14-15, 21, 54, 92

Solubility (continued):
of sodium chlorate, 92
in wood and grain alcohol, 57
Solubility curves, 53-55
Solutes, 50
Solutions:
concentration of, 51-52
frozen, 143
mass of, 22-23
saturated, 50
Solvent(s), 50
methanol, 58
Spectra:
analysis of, 116
definition of, 114
of the elements, 113-114
of helium, 190
of polonium, 190
rainbow, 114
Spectral analysis, 114-118
Spectral lines, 114-118
Spectroscope, grating, 114
Spectroscopy, 116
Sphere-gas machine, 154-155, 167-168
Stronium, spectral lines of, 115
Substances:
identifying, 44-46
properties of, 2-3
pure, 84-87
separation of, 67-87
with similar properties, 45
Sugar, properties of, 58
Sulfur, heated with copper, 22
Sulfuric acid, 59-61
Sunlight, analysis of, 116
Synthesis, 95-100

T

Tars, 56, 73
Temperature(s):
and density, 43
low, 82-84
and molecular speed, 167-170, 178
and radioactivity, 122
room, 53
and solubility, 53-54
Thermal expansion, 43
of gases, 170
at high pressures, 170
of liquids, 170-172
of solids, 173-175
Thermometer:
calibration of, 83
gas, 83
liquid, 83
Thickness:
of oleic acid film, 186-188
of thin layer, 183-184
Thin layer, thickness of, 183-184

U

Unified atomic mass unit (u), 199
Unit volume, mass of, 36
Uranium, radioactivity of, 121

V

Vapor, in fractional distillation, 69
Vibration, 177
Vitriol, oil of, 59
Volta, Alessandro, 112
Volume, 7-16
of an atom, 198
of a cube, 198
of a cylinder, 186
and displacement of water, 13-15
international standard of, 9
as a measure of matter, 15-16
measurement of 7-10, 183-188
and pressure of gases, 157-165
of a rectangular solid, 9, 183-184
units of, 9

W

Water:
and concentration, 50
decomposition of, 93-95
displacement of, 13-15
produced in burning, 61
synthesis of, 95-97
Wine, 57
Wood:
alcohol, 56-57
distillation of, 3-6, 56

Z

Zeros in decimal measurements, 12, 38
Zinc chloride, 98-100, 107